Lecture Notes in Computer Scien

T0250719

Commenced Publication in 1973
Founding and Former Series Editors:
Gerhard Goos, Juris Hartmanis, and Jan van Leeuwen

Paul Havinga   Maria Lijding
Nirvana Meratnia   Maarten Wegdam (Eds.)

Smart Sensing and Context

First European Conference, EuroSSC 2006
Enschede, Netherlands, October 25-27, 2006
Proceedings

 Springer

Volume Editors

Paul Havinga
Maria Lijding
Nirvana Meratnia

University of Twente
Department of Computer Science
P.O.Box 217, 7500 AE Enschede, The Netherlands
E-mail: {p.j.m.havinga; m.e.m.lijding; n.meratnia}@ewi.utwente.nl

Maarten Wegdam
Lucent - Bell Labs
Capitool 5, 7521 PL Enschede, The Netherlands
E-mail: wegdam@lucent.com

Library of Congress Control Number: 2006934577

CR Subject Classification (1998): H.3, H.4, C.2, H.5, F.2

LNCS Sublibrary: SL 5 – Computer Communication Networks and
Telecommunications

ISSN   0302-9743
ISBN-10   3-540-47842-6 Springer Berlin Heidelberg New York
ISBN-13   978-3-540-47842-3 Springer Berlin Heidelberg New York

Springer is a part of Springer Science+Business Media

springer.com

© Springer-Verlag Berlin Heidelberg 2006
Printed in Germany

Typesetting: Camera-ready by author, data conversion by Scientific Publishing Services, Chennai, India
Printed on acid-free paper   SPIN: 11907503   06/3142   5 4 3 2 1 0

Preface

This volume contains the papers and posters selected for presentation at the First European Conference on Smart Sensing and Context (EuroSSC 2006) in Enschede, The Netherlands. EuroSSC 2006 was the first conference of a series aiming at bringing together designers, engineers and researchers to explore two complementary viewpoints:

- A device-centric, technology-driven view: concerning intelligent sensors, sensor networks and information processing for a new generation of networked devices and environments.
- A service-centric, user-driven view: exploring architectures, techniques, and algorithms for context-aware and pro-active applications made possible by the diffusion of ambient communication, cooperating objects, and interaction technologies.

These subjects are active and relevant research areas in themselves, and there are several conferences that address them separately. EuroSSC 2006, however, considered them both, and especially the symbiosis between them, which we expect to result in very inspiring and interesting discussions, as well as new research ideas on how to combine them.

The conference was organized in single tracks covering various issues ranging from intelligent sensors, sensor networks, context management and context awareness, and privacy, to applications and test beds. Organizing a conference for the first time requires lots of preparations, such as finding a publisher, sponsoring organizations, and TPC members and most importantly attracting potential submitters. Fortunately, the amount and quality of the submissions were such that we were in the luxurious position to be able to accept only high quality and relevant papers. The conference attracted world wide attention and submissions came from five continents. A total of 15 accepted full papers and 14 accepted posters came from Asia, North America and Europe. All full papers underwent peer blind reviewing by at least three reviewers, and were judged based on their novelty, technical quality, account of prior work, readability and relevance. The acceptance rate for full papers was 27%. Poster descriptions were reviewed by two referees and accepted posters appear as short papers in the proceedings.

The technical program was complemented by interesting keynotes from Anind Dey and Kevin Warwick, titled *End-User Control in the Smart Home*, and *Upgrading Humans' Technical Realities and New Morals*, respectively. Besides papers, posters, and keynotes, the technical program also included a debate on the social and economical impact of ambient technology.

The EuroSSC 2006 conference was technically co-sponsored by the IEEE Communications Society and supported by the Ministry of Economic Affairs of the Netherlands through the Smart Surroundings, Freeband, and MultimediaN projects, the European IST funded e-SENSE project, Ambient Systems B.V., CTIT, and was organized in cooperation with EuSAI and ISSNIP.

Apart from the above listed organizations and projects, we would also like to express our gratitude to the many individuals who contributed to organizing EuroSSC 2006 and offering technical and administrative support. Specifically, we want to acknowledge the TPC members, additional referees, LNCS staff, and keynote speakers for their contributions.

August 2006   Paul Havinga
                                                     Maria Lijding
                                                   Nirvana Meratnia
                                                  Maarten Wegdam

Organization

Program Co-chairs

Paul Havinga   University of Twente, The Netherlands
Maria Lijding   University of Twente, The Netherlands
Nirvana Meratnia   University of Twente, The Netherlands
Maarten Wegdam   Lucent - Bell Labs, The Netherlands

Steering Committee

Emile Aarts   Philips, The Netherlands
Thijs Krol   University of Twente, The Netherlands
Sape Mullender   Lucent - Bell Labs, USA
M. Palaniswami   University of Melbourne, Australia

Program Committee

Peter Apers   CTIT, The Netherlands
Stefan Arbanowski   Fraunhofer FOKUS, Germany
Sebnem Baydere   Yeditepe Univ., Turkey
Hartmut Benz   WMC, The Netherlands
Srdjan Capkun   Tech. Univ. of Denmark, Denmark
Simon Dobson   Univ. College Dublin, Ireland
Henk Eertink   Telematica Instituut, The Netherlands
Berry Eggen   Eindhoven Univ. of Tech., The Netherlands
Ling Feng   Univ. of Twente, The Netherlands
Aart van Halteren   Univ. of Twente, The Netherlands
Sonia Heemstra de Groot   WMC, The Netherlands
Geert Heijenk   Univ. of Twente, The Netherlands
Hermie Hermens   Roessingh R&D, The Netherlands
Pierre Jansen   Univ. of Twente, The Netherlands
Mika Klemettinen   Nokia, Finland
Gerd Kortuem   Univ. of Lancaster, UK
Koen Langendoen   Delft Univ. of Technology, The Netherlands
Rodger Lea   Univ. of British Columbia, Canada
Peter Leijdekkers   Univ. of Technology Sydney, Australia
Qing Li   City Univ. of Hong Kong, Hong Kong, China
Pedro Marron   Univ. of Stuttgart, Germany
Slaven Marusic   Univ. of New South Wales, Australia
Ignas Niemegeers   Delft Univ. of Technology, The Netherlands

Stephan Olariu	Old Dominion Univ., USA
Anibal Ollero	Univ. of Seville, Spain
Zhiyong Peng	Wuhan Univ., China
Christopher Roadknight	BT Labs, UK
Ilja Radusch	Fraunhofer FOKUS, Germany
Kay Römer	Federal Institute of Technology, Switzerland
Hans Scholten	Univ. of Twente, The Netherlands
Mihail L. Sichitiu	North Carolina State Univ., USA
Tod Sizer	Lucent - Bell Labs, USA
Hong va Leong	Polytechnic Univ., Hong Kong, China
Michele Zorzi	Univ. of Padova, Italy

Additional Referees

C. Fischer	L. Evers
K. Wac	R. Gemesi
H. Mei	J. Wu
M. Setten	K. Muthukrishnan
C. Jacob	M. Marin-Perianu
D. Linner	S. Chatterjea
M. Kleise	R. Marin-Perianu
N. Bui	H. Teunissen
A. Zanella	K. Sheikh
F. De Pellegrini	U. Bischoff
D. Miorandi	T. Broens
K. Wah Chow	P. Pawar
H. Liu	G. Halkes
M. Kamilova	S. Dulman
C. Räck	O. Durmaz Incel
L. Hoesel	R. Neisse

Sponsoring Institutions

IEEE Communications Society
BSIK Funded Smart Surroundings, Freeband, and MultimediaN projects
European IST funded e-SENSE project
Ambient Systems B.V.
CTIT

Table of Contents

Intelligent Sensors and Sensor Network

Multi-channel Support for Dense Wireless Sensor Networking 1
Ozlem Durmaz Incel, Stefan Dulman, Pierre Jansen

Data Aggregation for Target Tracking in Wireless Sensor Networks 15
Caspar Lageweg, Johan Janssen, Maarten Ditzel

A Zone-Based Clustering Method for Ubiquitous Robots Based
on Wireless Sensor Networks 25
Kyungmi Kim, Hyunsook Kim, Young Choi, Sukgyu Lee, Kijun Han

Context Awareness and Architectures

A Simulation Study of Integrated Service Discovery 39
Gertjan P. Halkes, Aline Baggio, Koen G. Langendoen

Context Dissemination and Aggregation for Ambient Networks: Jini
Based Prototype 54
*Kazimierz Bałos, Tomasz Szydło, Robert Szymacha,
Krzysztof Zieliński*

Discovery and Composition of Services for Context-Aware Systems 67
Cristian Hesselman, Andrew Tokmakoff, Pravin Pawar, Sorin Iacob

Infrastructural Support for Dynamic Context Bindings 82
Tom Broens, Aart van Halteren, Marten van Sinderen

Adding Context Awareness to C# 98
Anca Rarau, Ioan Salomie

Toward Wide Area Interaction with Ubiquitous Computing
Environments 113
Michael Blackstock, Rodger Lea, Charles Krasic

Maintaining a World Model in a Location-Aware Smart Space 128
R.K. Harle

Privacy, Application and Test Beds

Shadow: A Middleware in Pervasive Computing Environment
for User Controllable Privacy Protection 143
    Wentian Lu, Jun Li, Xianping Tao, Xiaoxing Ma, Jian Lu

Auditing and Inference Control for Privacy Preservation in Uncertain
Environments 159
    Xiangdong An, Dawn Jutla, Nick Cercone

Developing a Context-Aware System for Providing Intelligent Robot
Services. 174
    Chung-Seong Hong, Joonmyun Cho, Kang-Woo Lee, Young-Ho Suh,
    Hyun Kim, Hyun-Chan Lee

Music for My Mood: A Music Recommendation System Based
on Context Reasoning 190
    Jae Sik Lee, Jin Chun Lee

WLAN Location-Aware Application Based on Accumulated Orientation
Strength Algorithm 204
    I-En Liao, Kuo-Fong Kao, Ke-An Chen

Posters: Short Papers

Context Delivery in Ad Hoc Networks Using Enhanced Gossiping
Algorithms. 218
    Syarulnaziah Anawar, Lorcan Coyle, Simon Dobson, Paddy Nixon

An Attribute-Based Naming Architecture for Wireless Sensor Networks
Using a Virtual Counterpart Overlay Network 222
    Eui-Hyun Jung, Yong-Pyo Kim, Yong-Jin Park, Seong-Yun Cho,
    Su-Young Han

A Sensor Platform for Sentient Transportation Research 226
    Jonathan J. Davies, David N. Cottingham, Brian D. Jones

Attention-Based Information Composition for Multicontext-Aware
Recommendation in Ubiquitous Computing 230
    Sungrim Kim, Joonhee Kwon

Context-Aware Trust Domains. 234
    Ricardo Neisse, Maarten Wegdam, Marten van Sinderen

An Evaluation Framework for Disseminating Context Information
with Gossiping 238
  Graham Williamson, Graeme Stevenson, Steve Neely,
  Simon Dobson, Paddy Nixon

Dynamic Bayesian Networks for Visual Surveillance with Distributed
Cameras 240
  Wojciech Zajdel, A. Taylan Cemgil, Ben J.A. Kröse

Embedded Intelligence: Enabling In-Situ Power Management
for Wireless Sensor Networks 244
  Rui Ma, Gregory M.P. O'Hare, Michael J. O'Grady

Proximity Sensing Using IEEE 802.15.4 Radios 248
  Mark Lowton, James Brown, Joe Finney, Gerd Kortuem

Towards Hovering Information 250
  Alfredo Villalba, Dimitri Konstantas

Balancing Smartness and Privacy for the Ambient Intelligence 255
  Harold van Heerde, Nicolas Anciaux, Ling Feng, Peter M.G. Apers

Energy Conservation with EDFI Scheduling 259
  Tjerk Bijlsma, Pierre Jansen

RuleCaster: A Programming System for Wireless Sensor Networks 262
  Urs Bischoff, Gerd Kortuem

Losing Control in Pro-active Home Environments 264
  Martijn H. Vastenburg

Author Index 267

Multi-channel Support for Dense Wireless Sensor Networking

Ozlem Durmaz Incel, Stefan Dulman, and Pierre Jansen

University of Twente, P.O. Box 217, 7500 AE Enschede, The Netherlands
{durmazo, dulman, jansen}@cs.utwente.nl

Abstract. Currently, most wireless sensor network applications assume the presence of single-channel Medium Access Control (MAC) protocols. When sensor nodes are densely deployed, single-channel MAC protocols may be inadequate due to the higher demand for the limited bandwidth. To overcome this drawback, we propose multiple channel support for improving the performance. Our method allows the nodes to utilize new frequency channels which results in the significant increase on the number of nodes that are granted access to the wireless medium. The method requires only one half-duplex transceiver per node, which is capable of sending and receiving over distinguished frequency channels. Simulation results show that, method successfully utilizes multiple channels and increases the performance proportional to the number of available frequencies for an example single-channel MAC protocol, LMAC.

1 Introduction

Wireless Sensor Networks (WSN) [1], is an evolving technology that is the fundament of various ubiquitous applications. WSN is embedded into the real world and enables monitoring, inspection and analysis of unknown, untested environments with battery operated, tiny sensor devices. Sensor nodes are designed to collect sensor data about the context and to transmit the readings by wireless communication.

With the growing interest, in the near future, WSN will be deployed everywhere in large numbers, which may be of the order of hundreds or thousands and even more [2]. The underlying protocols must be able to deal with these numbers of nodes. In a dense network, demand is higher for the limited bandwidth. This results in less chance to access the wireless medium due to higher contention in a dense neighborhood[1]. Besides the large numbers, limited channel capacity and the influence of interference due to external networks or electronic devices, that share the same parts of the spectrum, will result in a competitive communication environment.

---

[1] We define the neighborhood of a node as the set of nodes which are located within the node's transmission range. We consider a dense network where a node -on the average- has more than 50 nodes in its neighborhood.

P. Havinga et al. (Eds.): EUROSSC 2006, LNCS 4272, pp. 1–14, 2006.

The important reason for this competition is that sensor nodes share a single channel[2]. If the transceiver equipment used for wireless communication is able to operate on multiple non-overlapping channels rather than a single channel, multiple transmissions can take place on the wireless medium without disturbing each other. This leads to lower contention, less collisions and retransmissions. Today's transceiver hardware, which is used for sensor nodes, supports operation on multiple frequencies. For example, the radio used by Ambient μNode [3] and CC2420 radio [4] for MICAz and Telos sensor nodes can be tuned to different channels.

We consider the LMAC protocol [5],[6] as an example to show the inefficiency of single-channel MAC protocols in densely deployed sensor networks. LMAC is a light-weight and energy efficient MAC protocol proposed for WSN. It is based on the concept of scheduled access to the wireless medium. Each node controls a timeslot to transmit its data. Timeslots are selected in a distributed, self-configuring way. Further details of the LMAC protocol will be given in Section 3.

Besides its advantages, LMAC's operation depends on the number of timeslots, and in turn on the density of the neighborhood. When all timeslots are exhausted, the node may not be able to access the wireless medium and remains in its initialization state to find an empty timeslot. As the neighborhood gets denser, the number of required timeslots grows rapidly. Therefore, we need a mechanism that reduces the contention in the neighborhood and allows a node to control a timeslot for transmitting its data.

We propose to multiplex the timeslots with the frequency domain for using the spectrum more efficiently. Note that this approach does not use different transceivers; instead one half-duplex transceiver is sufficient. The proposed method allows the nodes to switch their transceivers on new channels on-demand, if the network reaches a density limit. One may argue that switching to different channels by a half-duplex transceiver will cause disconnections in the network. However, our method optimizes connectivity, i.e., connects as many neighbors as possible via multiple channels.

The method is composed of two phases. In the first phase, nodes select timeslots according to the single-channel LMAC rules. In the second phase, nodes select timeslots and also channels to communicate on. The LMAC protocol ensures that a timeslot is only reused after at least 2-hops. Thus, the number of timeslots that a node can select is not only limited by the number of 1-hop neighbors but also by 2-hop neighbors. However, if multiple channels are available, nodes are allowed to select those timeslots that are occupied by their 2-hop neighbors on different channels. The second phase is based on this idea of utilizing different channels by allowing concurrent transmissions.

The rest of the paper is organized as follows: Section 2 summarizes related work. Section 3 introduces the LMAC protocol. Section 4 describes the multi-channel support for LMAC protocol. Section 5 reveals the performance of the method by experimental simulations. Section 6 discusses some concluding remarks and suggestions for future work.

---

[2] A channel is defined to be a frequency range over which two nodes communicate. We will use the terms "channel" and "frequency" interchangeably in the text.

2 Related Work

The channel assignment problem and multi-channel MAC protocols in wireless networks have been extensively studied. Usage of multiple channels in multi-hop ad hoc networks has been shown to increase the throughput considerably, by allowing concurrent transmissions on different non-overlapping channels [7], [8], [9], [10], [11], [12]. Details about the algorithms and comparisons are explained by Mo et al,. [13].

When we look into WSN domain, characteristics are quite different from the ad-hoc networks. A typical sensor device is usually equipped with a single half-duplex radio transceiver, which can not transmit and receive simultaneously, but can work on different channels separately. On the other hand, traditional wireless ad hoc networks usually assume more powerful radio hardware and multiple transceivers per node. For instance, typical bandwidth used by WSN is usually very limited (e.g., 50Kbps). Zhou et al. [14] showed why multi-channel MAC protocols which are based on IEEE 802.11 are not suitable for WSN with respect to the packet size, RTS/CTS mechanism and limited bandwidth. We show why single-channel MAC protocols are not efficient in densely deployed sensor networks, for an example single-channel MAC protocol.

Zhou et al., [14] recently introduced the MMSN multi-frequency MAC protocol especially designed for WSN. MMSN consists of two aspects: frequency assignment and medium access. In frequency assignment, each node is assigned a frequency for data reception. Hence, a node intending to transmit should know about the receiver's frequency. Broadcast packets are transmitted on a dedicated channel. Medium access is a combination of contention and scheduled operation. Our method is not a complete MAC protocol proposal, but it provides multiple channel support for an example single-channel MAC protocol: Timeslots are multiplexed with frequency domain on demand, if the number of timeslots in the neighborhood is exhausted on a single channel.

IEEE 802.15.4 standard [15] also provides multiple channels for Personal Area Networks (PAN). The idea is to use non-conflicting channels to identify different PAN's. This is different than our approach where we introduce multiple channels in a dense network when the demand for the limited bandwidth is higher.

3 The LMAC Protocol

LMAC [5] is an energy-efficient medium access protocol designed for WSN. The protocol enables the communicating entities to access the wireless medium on a time-scheduled basis over a single frequency channel. Time-scheduled method has a natural advantage of collision free medium access, which avoids wasting energy and time.

Like other time-scheduled MAC algorithms, LMAC also considers time to be divided into slots which are further organized into periodic frames. A node with the intention to transmit can take control of a timeslot. A node transmits a control message at the beginning of its timeslot to address the receiver nodes.

Neighbor nodes must always listen at the beginning of a timeslot which contains information about the intended receivers, synchronization and the current timeslot. If neighbor nodes discover that they are not the intended receivers they turn off their power-consuming transceivers.

The timeslot selection mechanism in LMAC is fully distributed, thus needs no base-stations or central authorities to decide and allocate the timeslots to the nodes. In addition, the multi-hop nature of the WSN allows the timeslots to be reused.

For timeslot selection, the nodes use an algorithm based on local information only. Each node maintains a vector of length equal to the number of timeslots. This vector is used for storing the occupied slots within the 2-hop neighborhood. Initially, the vector is cleared. Nodes transmit information in the control message about those timeslots that the node considers to be occupied by itself and its 1-hop neighbors. When a packet is received, the logical OR operation is executed to update the information about the occupied slots in the neighborhood and the information is stored in the vector. If a node is not yet controlling any timeslot, it selects one from the free slots. This method ensures that a timeslot is only reused after at least 2-hops. The distributed algorithm for timeslot selection is shown in Figure 1. The node marked with "?" is searching for a timeslot and other nodes control the timeslots they are marked by. It receives the occupied slots information from the neighbors, executes the OR operation and finds timeslot 7 as free and grabs it.

When there are no more free slots (i.e. in a dense neighborhood), the node remains in its initialization state, periodically monitoring frames to find an empty timeslot. Reserving a timeslot for each node or increasing the number of timeslots

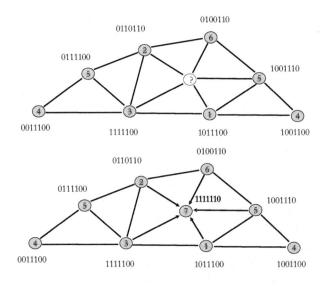

Fig. 1. LMAC Protocol: Timeslot selection

in the network may be a possible solution. However, this would increase the latency of communication and time of waiting, before nodes get the opportunity to transmit. For this reason, the frame interval should be kept as short as possible and reused as much as possible. Currently, in a typical LMAC implementation, frame length is 1 second and a frame consists of 32 timeslots.

4   Multi-channel Support for LMAC

To overcome the deficiency of single-channel LMAC protocol in dense networks, we propose an algorithm which utilizes multiple channels. Besides, the algorithm optimizes connectivity, i.e., connects as many neighbors as possible.

The algorithms is based on local information only: it does not need central authorities to decide and allocate the frequencies. For instance, a central solution for multiple channel allocation -which is also used in cellular networks or DECT- would be to let a base-station (for example a sink node which is a gateway between the users of the network and the network itself in WSN) assign the channels. The sink node could assign frequencies to its 1-hop neighbors, and frequency information can be broadcast to the remaining nodes, in a multi-hop fashion. The nodes that are receiving the broadcast messages with frequency information, switch to the associated frequency. In this approach, the ultimate view of the network is partitions communicating on different frequencies. Nodes are only aware of their neighbors which are using the same channel. If two sensor nodes are within each other's neighborhood but they are not connected neighbors, the data packets that are destined for each other have to travel all the way from the sender to the sink node and from sink node to the destination node. This method would require a lot of messages to be relayed, and cause waste of energy and latency.

Before describing the algorithm in detail, we summarize the design issues and assumptions.

4.1   Design Issues and Assumptions

- N non-overlapping frequency channels are available. Nodes are aware of the number and frequency range of the channels.
- All the nodes are communicating in the basic channel (single channel) at the beginning. If timeslots are exhausted in a node's neighborhood on the basic channel, new channels are introduced.
- The switching delay from one channel to another can be neglected, e.g., for the transceiver of Ambient μNode sensor node platform, 650μsec is much less than a typical timeslot duration, 31.25msec.
- Each node has one radio interface which is a half-duplex transceiver. A node cannot both transmit and receive at the same time, simultaneously.
- To establish multi-hop time synchronization ([16], [17]), every node uses its parent node to synchronize to every frame. A node can be a parent of another node if it's closer to the sink.

4.2 Functional Description

The algorithm is composed of 2 phases. In the first phase, nodes select the times-lots according to the single-channel LMAC rules. In the second phase, channel selection takes place. The second phase is based on 2-hop neighborhood infor-mation. The LMAC protocol ensures that a timeslot is only reused after at least 2-hops. Thus, the number of timeslots that a node can select is not only limited by the number of 1-hop neighbors but also by 2-hop neighbors. Moreover, the simulation results (Section 5) have shown that on the average, more than 30% of the occupied slots of a node are claimed by its 2nd hop neighbors. A node an-nounces these 2-hop slots as free for its neighbors. However, the slots cannot be grabbed by a slotless node since all timeslots are occupied in its neighborhood. A slotless node cannot use the announced free slots on the same channel but can do so on a different channel. Therefore it monitors the announced free slots and marks the announcing node as a potential bridge. We call this node as a bridge node since it connects the slotless node to the rest of the network. A slotless node selects a bridge among the potential bridges and negotiates with the bridge on an appropriate slot/frequency pair. This negotiation keeps the network connected and the new joining node does not disturb the current established connectivity.

Figure 2 shows the idea of the second phase. In the figure, the slotless node is marked by a "?". Other nodes control the timeslots they are marked by. Node 3 announces its occupied vector as "1011" where node 4 is announcing as "0111". All timeslots in ?'s neighborhood are occupied (1011 OR 0111 = 1111). If a single-channel LMAC protocol was used, the node could not grab a slot in this situation. If multiple channels are available, after receiving this information, the slotless node views node 3 on slot 2 and node 4 on slot 1 as potential bridges. We explain the algorithm in the next sections based on the example given in Figure 2.

Bridge and non-interfering Channels Discovery. In the first phase of the algorithm, nodes are communicating on the basic channel and they are only aware of their neighborhood on that channel. However, before a slotless node started the second phase of searching for potential bridge nodes, other slotless nodes may have already connected to the network on different channels. In order to be aware of all potential bridges operating on different channels, a slotless node scans different channels for bridge node discovery. This process also helps the node to discover all the occupied timeslots in different frequencies before deciding on a non-interfering frequency/timeslot pair.

A slotless node creates two matrices about the collected information. One is for identifying the occupied slot/frequency pairs in the neighborhood. For the example shown in Figure 2, the slotless node creates the matrices shown in Figure 3. Here, number of timeslots is 4 and number of frequencies is 3. In the occupied matrix, all the slots on the basic channel (frequency 0) are occupied where other channels are marked as free. In the free matrix, the slotless node records node 3 on slot 2 and and node 4 on slot 1 as potential bridges in any of the available frequencies other than the basic channel. After constructing the matrices, the slotless node extracts the required information about potential bridges. A bridge is selected due to some function from the set of potential bridges, such as the signal strength, degree of connectivity, battery level, etc.

Channel Negotiation. For the bridge node to be aware of a slotless node requesting negotiation, the slotless node should be able to send its request. However, it does not have a timeslot to transmit. One option for negotiation would be to take place on a dedicated control channel. Potential bridge nodes would switch to the control channel on slots which they announce to be free and listen for the requests. However, this would be very costly for the nodes in terms of energy if they have most of their timeslots as free. They would have to keep their transceiver always on, and probably receive no request. Instead, when a potential bridge node does not have data to send, it sends a notification for those slotless nodes which are interested in negotiation, during its control message (Section 3). In the rest of the timeslot, during data section the slotless node sends its request to the bridge node. This request includes the channel information on which the slotless node intends to communicate with the bridge. Here, a question may arise what if there are more slotless nodes which are intending to negotiate with this bridge node. To prevent collisions, nodes send the request based on a contention mechanism.

After getting the request, the bridge node also checks to investigate whether there are any conflicting transmissions on that slot/frequency pair in its neighborhood. It acknowledges the slotless node during its timeslot in the next frame if there is no conflict. After starting transmissions, still there may be collisions or interference with some other nodes. In this case, the agreement is canceled by the both parties and the requesting slotless node restarts the same process. For the example shown in Figure 2, let's assume that node ? has agreed with

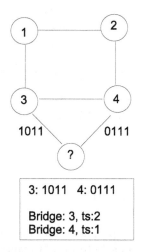

Fig. 2. Operation of the algorithm

Occupied Matrix

Timeslots/Frequencies	0	1	2
1	occupied	-	-
2	occupied	-	-
3	Node 3	-	-
4	Node 4	-	-

Free Matrix

Timeslots/Frequencies	0	1	2
1	occupied	Node 4	Node 4
2	occupied	Node 3	Node 3
3	occupied	-	-
4	occupied	-	-

Fig. 3. Matrices discovered by node "?"

Node 1	1	2	3	4
Frequency	0	0	0	-
Neighbor	myslot	Node2	Node3	-
Node 2	1	2	3	4
Frequency	0	0	-	0
Neighbor	Node1	myslot	-	Node4
Node 3	1	2	3	4
Frequency	0	1	0	0
Neighbor	Node1	Node ?	Node3	-
Node 4	1	2	3	4
Frequency	-	0	0	0
Neighbor	-	Node2	Node3	myslot
Node ?	1	2	3	4
Frequency	-	1	0	-
Neighbor	-	myslot	Node3	-

Fig. 4. Occupied vectors with frequency information

node 3 on frequency 1 over slot 2. The occupied vectors for the nodes are shown in Figure 4.

After a slotless node has negotiated with a bridge node and is further controlling a timeslot, it can also play role as a potential bridge as well since it's already a part to the network. A sketch about execution model of multi channel LMAC is shown in Figure 5.

5   Performance Evaluation

In this section we present some experimental results concerning the inefficiency of LMAC protocol in dense networks, in terms of the number of active nodes (nodes that control a timeslot). Moreover, we investigate the overhead of using multiple channels. We have carried out simulations in the Omnet++ environment [18].

The aim of doing experiments is to prove that the concept of the multi-channel support for LMAC protocol is valid.

Fixed simulation parameters are tabulated in Table 1. Sensor nodes are deployed randomly within the terrain, and are assumed to be static during the simulation interval. A topology generator tool is used to deploy sensor nodes randomly (with a uniform distribution) within the given dimensions of terrain size. We create 5000 random topologies and for each simulation run, (whether for LMAC or multi-channel LMAC version) the same topology is used. By changing the number of nodes but keeping the terrain size and the transmission range fixed, we simulate different levels of neighbor density. We adopt the neighbor density calculation from Bulusu et al. [19], as:

$$\mu = (N \Pi R^2)/A \tag{1}$$

where N is the number of nodes, R is the transmission range and A is the size of the terrain. So densities for different number of nodes (50, 100, 150, 200, 250, 300, 350, 400, 450, 500, 550) are approximately 24, 48, 72, 96, 120, 144, 168, 192, 216, 240 and 264. Note that these numbers include only 1-hop densities and don't include 2nd hops. So, the neighbor densities in the example topologies are larger than these numbers and almost covering all of the nodes in the network.

```
controlled_slot = NO_VALUE
While (phase 1)
do
    if(current_slot == controlled_slot)
        transmit
    else
        receive
    if(controlled_slot == NO_VALUE)
        grab a timeslot
    update synchronization
    update current_slot
done
while (phase 2)
do
    if(controlled_slot == NO_VALUE)
        role = slotless
        discover potential bridge nodes
        select a bridge/frequency/slot
        send a negotiation request
        if(acknowledgment is received)
            controlled slot = selected slot
        else
            Select another bridge/frequency/slot
    else
        if(occupied_slots_local != full)
            role = potential_bridge
            if(negotiation request is received)
                check the proposed frequency during proposed slot
                if(free)
                    acknowledge in the next frame
                    switch to the proposed frequency during proposed slot

        if(current_slot == controlled_slot)
            Transmit
        else
            Receive
done
```

Fig. 5. Multi channel support for LMAC: execution model

Table 1. Simulation Parameters

Parameter Name	Value
Terrain Size	100*100 m^2
Transmission Range	40m
Sensor Node Deployment	Uniform
Mobility Characteristic	Static
Density of Neighborhood	24, 48, 72, 96
	120 144, 168, 196 nodes
Number of Time Slots	32 timeslots
MAC frame size	1 sec.
Number of Frequencies	8 (non-overlapping)
Number of Runs	5000

The traffic pattern used in the simulations is that every node broadcasts its occupied vector during its timeslot in its neighborhood. We assume error-free links. The simulation ends either when all the nodes control a timeslot or when there are no more timeslots that can be grabbed by a node. The simulation time limit is 100 seconds.

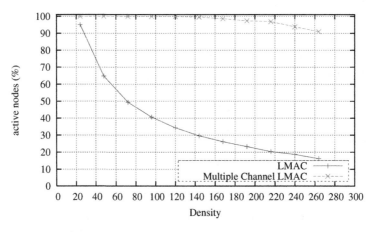

Fig. 6. Percentage of active nodes versus density

Firstly, we show how multiple channels increase the number of active nodes in the network. In Figure 6, the percentage of active nodes ([The nodes that have a timeslot/Number of nodes]*100) versus different number of densities is shown. When the network becomes denser, the percentage of active nodes decreases rapidly in LMAC. For 240 neighbors, it is below 20%. Consequently, LMAC suffers from density. If we compare the results of multi-channel LMAC with pure LMAC's results we see that the performance is above 90% even in the most dense topologies. This method ensures that nodes can communicate, i.e. have a timeslot in a dense neighborhood even the neighborhood is getting denser with the increasing number of nodes and the same terrain dimensions. Multi-channel LMAC increases the performance by a factor of number of available channels. However, the number of active nodes is limited by "*number of timeslots * number of frequencies*". This shows that approximately all the channels are successfully utilized. In conclusion, LMAC's performance is affected by the number of timeslots whereas multi-channel LMAC increases the performance by the number of non-overlapping channels available.

To design a good MAC protocol for WSN, the first attribute to be considered is the energy efficiency [20] since the sensor nodes are usually battery-powered and it's difficult to change or recharge batteries. The major source of energy waste is the collisions. In this set of experiments we have tested how energy-efficient our method is in terms of collisions. Another source of energy-waste is the control packet overhead. We test the overhead of the control packets sent during the negotiation process. The results are shown in Figures 7 and 8. The collisions are represented per node for a fair comparison instead of totals. Because in LMAC number of active nodes is less and total number of collisions is also less. Number of collisions per active node in multi-channel LMAC is much less than LMAC. Multi-channel LMAC introduces multiple channels and this results in less number of collisions and contentions for timeslots. Control packets are not present in LMAC while in multi-channel LMAC, number of control packets increases with increasing density but it is still within boundaries between 0 and 5 per active node. These results are also represented in terms of control packets per node over a simulation period. The results prove that multi-channel LMAC does not bring an overhead in terms of energy.

Another parameter which directly affects the execution of the method is the number of available 2-hop slots offered by a node to be used by nodes in different frequencies. Results are shown in Figure 9. It's seen that in even most of the crowded topologies, more than 30% (10/32)of the slots are announced as free and slotless nodes can negotiate for different frequencies on these slots.

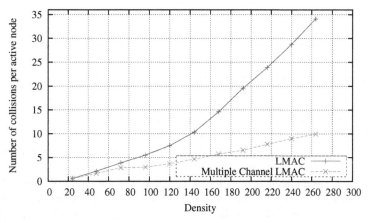

Fig. 7. Number of collisions per number of active nodes(%) versus density

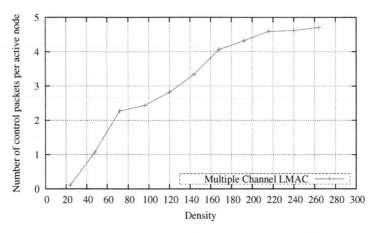

Fig. 8. Number of control packets per number of active nodes(%) versus density

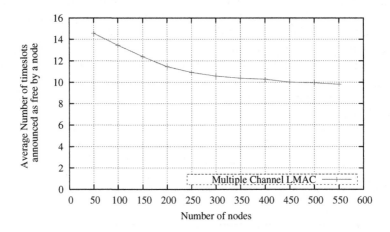

Fig. 9. Number of available slots announced by a node versus number of nodes

6   Conclusion and Future Work

In this future-looking work, we have addressed multiple channel support for WSN. The single-channel MAC protocols which grant access to the bandwidth may be inadequate in dense environments due to the higher demand and contention for the limited bandwidth. We propose a method that extends an example single-channel MAC protocol by multiplexing the time domain with the frequency domain. It allows the nodes to introduce new channels on-demand, if the network reaches a density limit. The method is composed of two phases. In the first phase, nodes try to select timeslots according to the single-channel LMAC rules. In the second phase, nodes that could not grab a timeslot in the first phase invite the neighbor nodes which are free to listen them on an agreed frequency channel and a timeslot.

The current state of the work proves that an example single-channel timeslotted MAC protocol "LMAC" suffers in dense environments and the performance can be increased by introducing the multiple frequency channels. Simulation results show that multiple channel support has the effect of increasing the performance by a factor of number available frequencies. Moreover, it does not bring an overhead in terms energy efficiency per node.

After the encouraging results, we will explore the adaptivity aspect of this solution and investigate the performance of multi-channels in other MAC protocols also with mobility scenarios. Currently, we are experimenting the method on a test-bed to have a clear idea about the number of non-overlapping channels available since the extent of spatial reuse is directly proportional to this number.

Acknowledgments

The authors gratefully acknowledge Sape Mullender for his help in revising the manuscript.

References

1. Akyildiz, I.F., Su, W., Sankarasubramaniam, Y., Cayirci, E.: A survey on sensor networks. IEEE Communications Magazine **40** (2002) 102–114
2. Gang Zhou, J.S., Son, S.: Crowded spectrum in wireless sensor networks. In: EmNets 2006: Proceedings of the Third Workshop on Embedded Networked Sensors. (2006)
3. http://www.ambient systems.net: Ambient-systems products line-up (2006)
4. http://www.chipcon.com: Cc2420, 2.4 ghz, ieee 802.15.4, zigbee-ready rf transceiver (2006)
5. van Hoesel, L., Nieberg, T., Wu, J., Havinga, P.: Prolonging the lifetime of wireless sensor networks by cross layer interaction. IEEE Wireless Communication Magazine **11** (2005) 78–86
6. Hoesel, L.v., Havinga, P.: A lightweight medium access protocol (LMAC) for wireless sensor networks. In: Proceedings of INSS04, Tokyo, Japan (2004)
7. Nasipuri, A., Zhuang, J., Das, S.: A multichannel csma-mac protocol for multihop wireless networks. In: Proceedings of IEEE Wireless Communications and Networking Conference (WCNC'99). (1999) 1402–1406
8. So, J., Vaidya, N.H.: Multi-channel mac for ad hoc networks: handling multichannel hidden terminals using a single transceiver. In: Proceedings of the 5th ACM international symposium on Mobile ad hoc networking and computing (MobiHoc '04). (2004) 222–233
9. Jain, N., Das, S., Nasipuri, A.: A multichannel csma mac protocol with receiver-based channel selection for multihop wireless networks. In: Proceedings of the IEEE IC3N. (2001) 432–439
10. Lin, C.Y.: A multi-channel mac protocol with power control for multi-hop mobile ad hoc networks. In: Proceedings of the 21st International Conference on Distributed Computing Systems (ICDCSW '01). (2001) 419

11. Roy, S., Das, A., Vijayakumar, R., Alazemi, H., Ma, H., Alotaibi, E.: Capacity scaling with multiple radios and multiple channels in wireless mesh networks. In: Proceedings of the First IEEE Workshop on Wireless Mesh Networks(WiMesh). (2005)

12. So, H.S.W., Walrand, J.: Design of a multi-channel mac protocol for ad hoc wireless networks. Technical Report UCB/EECS-2006-17, EECS Department, University of California, Berkeley (2006)

13. Mo, J., Sheung, H., So, W., Walrand, J.: Comparison of multichannel mac protocols. In: Proceedings of the 8-th ACM/IEEE International Symposium on Modeling, Analysis and Simulation of Wireless and Mobile Systems. (2005) 209–218

14. Zhou, G., Huang, C., Yan, T., He, T., Stankovic, J.A., Abdelzaher, T.F.: Mmsn: Multi-frequency media access control for wireless sensor networks. In: Proceedings of IEEE Infocom. (2006)

15. http://www.ieee802.org/15/pub/TG4.html: Ieee 802.15 wpan task group (2006)

16. Elson, J., Estrin, D.: Time synchronization for wireless sensor networks. In: IPDPS '01: Proceedings of the 15th International Parallel & Distributed Processing Symposium. (2001) 186

17. Elson, J., Roemer, K.: Wireless sensor networks: a new regime for time synchronization. SIGCOMM Comput. Commun. Rev. **33** (2003) 149–154

18. Varga, A.: Omnet++ community site (2006)

19. Estrin, D., Govindan, R., Heidemann, J., Kumar, S.: Next century challenges: scalable coordination in sensor networks. In: MobiCom '99: Proceedings of the 5th annual ACM/IEEE international conference on Mobile computing and networking. (1999) 263–270

20. Demirkol, I., Ersoy, C., Alagoz, F.: Mac protocols for wireless sensor networks: a survey. IEEE Communications Magazine **44** (2006) 115–121

Data Aggregation for Target Tracking in Wireless Sensor Networks

Caspar Lageweg, Johan Janssen, and Maarten Ditzel

TNO Defence, Security and Safety
P.O. Box 96864, 2509 JG Den Haag, The Netherlands
{caspar.lageweg, Johan.janssen, Maarten.ditzel}@tno.nl

Abstract. This paper presents the results of a study on the effects of data aggregation for target tracking in wireless sensor networks. In these networks energy, computing power and communication bandwidth are scarce. A novel approach towards data aggregation is proposed. It is tested in a simulation environment and compared with more straightforward methods. The results of the experiments clearly show the benefit of the new approach in terms of energy consumption and tracking accuracy.

1 Introduction

Wireless sensor networks [1,2] have experienced increasing attention in academic, industrial and military environments over the past few years. These networks promise an easy-to-deploy, easy-to-use and moreover, low-cost means to remotely monitor environments. Furthermore, sensing accuracy can be improved significantly by processing and combining collected data within the network itself. Finally, the network can be made robust to the failure of individual nodes, which ensures that the lifetime and proper operation of the network is not limited to the lifetime of one node in particular.

Applications, either envisioned or already realized, are generally related to the remote monitoring of a, possibly inaccessible or hostile, environment. Examples are an aqueous surveillance system for a drinking water reservoir [3], a wildlife habitat observation system [4], and a network to monitor the behavior of glaciers [5]. Additionally, numerous military applications are envisioned such as a battlefield data collection network as described in [6].

In all the aforementioned applications, the network consists of tens to thousands of tiny devices (e.g., see Figure 1). Each device carries one or more sensors and has limited signal processing and communication capabilities. Usually, the devices are powered by batteries and can thus only operate for a limited time period. Key to implementing a network with such devices is that energy, computing power and communication bandwidth are scarce.

Recent research efforts are dedicated towards operation of the network, focussing on energy efficient and robust communication schemes, reconfigurability, security, etc. [7,8,9]. Relatively less attention is payed to the actual goal of the networks: the collection and delivery of interesting information, extracted from

P. Havinga et al. (Eds.): EUROSSC 2006, LNCS 4272, pp. 15–24, 2006.
© Springer-Verlag Berlin Heidelberg 2006

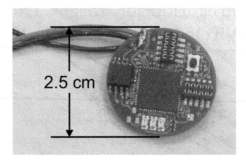

Fig. 1. Example of a wireless sensor node: the TNOdes

data that is gathered by one or more sensors. This paper presents the results of a short study on the tracking of objects in a network, focussing on the trade-off between the amount of communication in the network and the tracking accuracy.

The objective of this study is to reduce the the amount of messages sent within the network. This can be achieved by utilizing local processing and sensor data aggregation within the network. For our experiments we assumed a grid-based network through which an object of interest (a target) travels. As we are primarily interested in the effects of the aggregation strategies on the performance of the tracking, we assume ideal target detection and communication links. We investigated five experiments each having a different strategy for the transfer of sensor data to the sink. In order to evaluate the experiments two basic metrics are utilized: the total amount of messages sent in the network (both broadcasts and unicasts), and the quality of the position estimates expressed as the RMS difference of the estimated track and the original path of the observed target.

The remainder of this paper is organized as follows. Section 2 introduces tracking in wireless sensor networks and discusses the simulation framework and the experiments in more detail. Moreover, it describes the algorithm used in the experiments for target position estimation. Then, Section 3 summarizes and discusses the simulation results. Finally, Section 4 concludes the paper with some remarks.

2   Tracking in Wireless Sensor Networks

Wireless sensor networks typically consist of a large number of individual nodes with limited communication and observation ranges. An object of interest (target) that passes through such a network is only 'visible' to nodes within whose sensor range it is located. These nodes will then attempt to send the observation data in the form of messages to a collection point commonly referred to as the sink. Given that the communication range of sensor nodes is limited, sending a message from a node to the sink typically results in a series of hops through the network. Each of these hops results in the consumption of limited (battery) energy, which eventually results in failures within the network as nodes completely run out of energy (particularly in the vicinity of the sink where messages converge).

General object localization and tracking in sensor networks is actively researched and addressed in several papers, for example [10,11,12,13]. The authors focus on different aspects of tracking in sensor networks, such as real-time implementation aspects, sensor query systems, classification issues and node activation strategies. Data aggregation methods for target tracking as such, are not covered in detail.

2.1 Simulation Environment

For our experiments we utilized an event driven simulator that was developed using the OMNeT++ simulation framework [14]. The wireless sensor network simulated in our experiments consists of a 51×51 equidistant grid of sensor nodes, placed 100 meters apart. The communication range of the network is chosen such that nodes can only communicate with their orthogonal neighbors, as indicated by the dotted lines in Figure 2. It is assumed that the positions of the individual nodes are known by the nodes themselves. The sink is located within the center of the network. As a consequence all nodes can reach the sink with at most 50 hops.

The sensor range of each node is set such that a target within the network will be observed by multiple nodes. The simulated sensor nodes can detect range, which corresponds to a sensor node equipped with a simple radar [15]. The

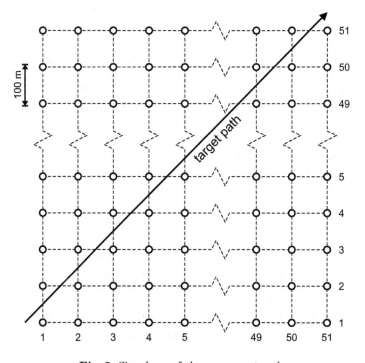

Fig. 2. Topology of the sensor network

nodes operate at 2 Hz, a common update interval for unattended ground sensors enabling the detection of pedestrians and vehicles at various speeds. The target (see Figure 2) travels through the sensor grid at constant speed with a slightly varying angle. Its average speed is approximately 4 km/h, which corresponds to a human's walking speed. All nodes are assumed to communicate error free. They can either communicate with each other through broadcasts (reaching all direct neighbors) or through unicast (peer-to-peer) communication, using idealized shortest path routing to reach the sink.

2.2   Experiments

In order to investigate the amount of messages sent out by nodes in the network we have conducted a set of five experiments. Each experiment uses a different strategy to send sensor data from observing nodes to the sink.

- *Experiment 1: reference*
  The first experiment does not use any local processing or data aggregation algorithm. During each time step, all nodes that observe the target report their findings to the sink using message hopping.

- *Experiment 2: delta messaging*
  The second experiment uses local processing. The first time step where a node observes a target, it reports its observation to the sink. From this time step forward it is assumed by the sink that the node can observe the target. When the target moves out of sensor range, the node reports it has lost the target. In other words, it only reports delta information (mutations in observations).

- *Experiment 3: local aggregation*
  The third experiment uses data aggregation. During each time step, all nodes that observe the target broadcast their findings to their neighbors. As part of this message they send an indication of the quality of the observation (i.e., the normalized strength of their sensor signal, see Section 2.3). Once all broadcasts have been sent, each node that observed the target compares the quality of its own observation with those reported in the broadcasts that it has received. If the quality of the node's own observation is better than those reported by others, it estimates the target's position by weighing each observing node's position with the quality of their observations. The estimates are reported to the sink.

- *Experiment 4: local aggregation and delta messaging*
  The fourth experiment combines the target position estimation described for experiment 3 with the delta messaging utilized in experiment 2. Nodes that have better observation quality than what is reported to them by others send a single message that informs the sink of the estimated position. While a node continues to have the best quality observation no other messages are sent to the sink. Once the node no longer has the best quality observation, (also when the target moves out of sensor range), a single message is sent to the sink to report this change.

– *Experiment 5: Fine-grained local aggregation and delta messaging*
  The fifth experiment generalizes experiments three and four. If a node has
  the best quality observation it sends a single message that informs the sink of
  the estimated position. As the target moves through the network, the nodes
  continue to estimate the targets position. If the distance between the last
  reported position and the last calculated position exceeds a given threshold,
  the new position is sent to the sink. Once a node no longer has the best
  quality observation, a single message is sent to the sink to report this change.
  Also when the target moves out of sensor range, a message is transmitted.

It should be noted that experiments 3 and 4 are in fact special cases of the
more general experiment 5, by setting the threshold to zero and infinity, respectively. The next section describes the position estimation algorithm based
on the collected sensor data. The algorithm is executed centrally at the sink in
experiments 1 and 2. In experiments 3, 4 and 5 it is run locally at the nodes.

2.3   Position Estimation

Once a target comes within sensor range of node i, its distance to the node is
measured. We express the quality of this observation as $q_i \in [0, 1]$, where $q_i = 0$
implies that the target is out of sensor range and $q_i = 1$ that the target is at the
same position as the sensor node. Each node i observing the target is assumed to
broadcast $\{x_i, q_i(t)\}$, where x_i is the position of the node, and $q_i(t)$ the quality
of the observation at time t, which is calculated using

$$q_i(t) = 1 - \frac{|x_i - x_o(t)|}{R_i}, \tag{1}$$

where $x_o(t)$ is the time dependent position of the observed target, and R_i the
maximum range of the node's sensor. If the quality of a node's own observation is
higher than that reported by others, the node calculates the sum of observation
qualities (it's own and those reported by others) which it uses to determine the
weight $w_i(t)$ of each observation as

$$w_i(t) = \frac{q_i(t)}{\sum_{n \in N} q_n(t)} \tag{2}$$

where N represents the set neighboring nodes and the node itself. Given the
weights $w_i(t)$ we calculate the target's estimated position $\hat{x}_o(t)$ as the weighted
sum of the observing nodes' positions

$$\hat{x}_o(t) = \sum_{n \in N} w_n(t) x_n. \tag{3}$$

3   Experimental Results

The communication that occurs during the experiments can be divided in three
categories of interest. First, one can count the total amount of messages that

Table 1. Communication metrics for experiments 1 to 5. Number of messages are listed per observation.

Experiment	Sink arrivals	Total messages	Broadcasts
1 - Reference	8.06	198.52	N/A
2 - Delta messaging	0.04	0.87	N/A
3 - Local aggregation	1.00	33.16	24.3%
4 - Local aggregation and delta messaging	0.02	8.45	95.4%
5 - Fine-grained aggregation and messaging	0.08	9.98	80.7%

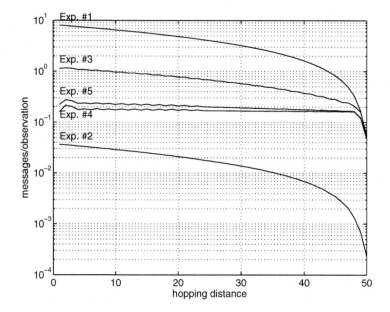

Fig. 3. Total number of messages per observation (unicast and broadcast) as function of the number of hops to the sink (hopping distance)

reach the sink and utilize this as an indication of the energy consumption of the nodes along the path from source to sink. Second, one can count the total amount of messages that are sent from one node to another (unicast and broadcast) as an indication of total energy consumption. Third, one can examine the fraction of broadcasted messages separately as broadcasts typically result in higher energy consumption due to the fact that more nodes are listening[1]. We have analyzed these three categories for the five experiments described previously and summarized the results in Table 1 and Figure 3.

Examining Table 1 one can observe the following. First, the total amount of messages in experiments 1 and 2 is about equal to the number of messages that

---

[1] In a typical sensor node transmitting and receiving require approximately the same energy. For example, in the TNOdes (see Figure 1, transmission and reception draw 16 mA and 12 mA, respectively.

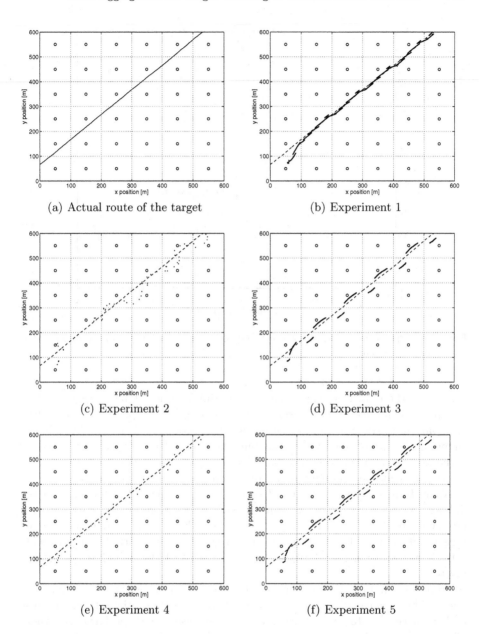

(a) Actual route of the target

(b) Experiment 1

(c) Experiment 2

(d) Experiment 3

(e) Experiment 4

(f) Experiment 5

Fig. 4. Position estimates received and processed by the sink. The dashed lines represent the actual route of the target. The circles depict the locations of the sensor nodes.

arrived to the sink multiplied with a factor of 25. When increasing or decreasing the size of the network, one can expect to see this factor increase and decrease proportionally as the average distance of the nodes to the sink varies. Second,

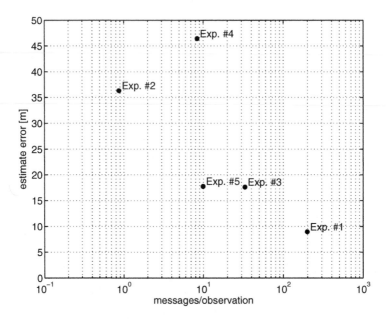

Fig. 5. Cost-performance trade-offs: estimation error versus the total number of messages per experiment

examining the data from experiments 3, 4, and 5 while taking experiment 1 as a basis for comparison, one observes that each node that previously sent their observations to the sink now broadcasts its observations to their neighbors. When changing the size of the network, the number of broadcasts will remain constant if the average number of nodes that observe the target remains unchanged (in our experiments on average 8 nodes observed the target simultaneously).

Apart from the cost in terms of transmitted messages, the influence of the different aggregation strategies is analyzed. The position estimates for each experiment are depicted in Figure 4. Figure 4(a) depicts the actual route of the target and Figure 4(b) to Figure 4(f) plot the positions reported in experiments 1 to 5, respectively. For clarity only a small area of the actual sensor grid is plotted.

From these figures, performance metrics can be can be derived for each experiment. They are calculated as the RMS difference between the estimated paths and the actual route of the target. Figure 5 shows the cost-performance trade-off for each experiment. The cost is expressed as the total number of messages sent per observation during the simulation. It is clearly seen that experiment 1 has the best performance (as can be expected, as all data is available at the sink), however it also has the highest cost. Delta messaging (experiment 2 and 4), although cheap, has the worst performance. Noteworthy is the large gain in performance between experiment 4 and 5, at the expense of only little extra communication.

4   Conclusions and Future Work

This paper discusses the results of a study on the effects of data aggregation for target tracking in wireless sensor networks. Various novel aggregation strategies have been implemented and analyzed, using a event driven simulation environment. For all experiments we used a target position estimation algorithm described in this paper, either run centrally at the sink, or locally at the nodes. The amount of communication in each experiment is analyzed as it largely accounts for the energy consumption in wireless sensor networks. It is compared with the performance of the data aggregation strategies.

The results from the experiments clearly show a relevant trade-off between the amount of communication and the performance of the position estimation. Moreover, the fine-grained local aggregation combined with delta messaging used in the fifth experiment enables us to tune the cost-performance trade-off to the user's requirements.

Current activities are focussed on implementing the various strategies in a real-world test-bed. Moreover, simulation framework and the aggregation strategies are extended to model more realistic scenarios, including non-uniform sensor grids and non-ideal target detections. In addition we are working on extra features such as early false target suppression and multiple track support.

References

1. D. Estrin, R. Govindan, J. Heidemann, and S. Kumar, "Next century challenges: scalable coordination in sensor networks," in *ACM/IEEE international conference on Mobile computing and networking*, pp. 263–270, ACM Press, 1999.
2. G. J. Pottie and W. J. Kaiser, "Wireless integrated network sensors," *Commun. ACM*, vol. 43, no. 5, pp. 51–58, 2000.
3. X. Yang, K. G. Ong, W. R. Dreschel, K. Zeng, C. S. Mungle, and C. A. Grimes, "Design of a wireless sensor network for long-term, in-situ monitoring of an aqueous environment," *Sensors*, vol. 2, pp. 436–472, November 2002.
4. A. Mainwaring, D. Culler, J. Polastre, R. Szewczyk, and J. Anderson, "Wireless sensor networks for habitat monitoring," in *International workshop on Wireless sensor networks and applications*, pp. 88–97, ACM Press, 2002.
5. K. Martinez, J. K. Hart, and R. Ong, "Environmental sensor networks," *Computer*, vol. 37, pp. 50–56, August 2004.
6. F. Ye, H. Luo, J. Cheng, S. Lu, and L. Zhang, "A two-tier data dissemination model for large-scale wireless sensor networks," in *International conference on Mobile computing and networking*, pp. 148–159, ACM Press, 2002.
7. T. van Dam and K. Langendoen, "An adaptive energy-efficient MAC protocol for wireless sensor networks," in *International conference on Embedded networked sensor systems*, pp. 171–180, ACM Press, 2003.
8. W. Ye, J. Heidemann, and D. Estrin, "An energy efficient MAC protocol for wireless sensor networks," in *Conference of the IEEE Computer and Communications Societies (INFOCOM)*, vol. 3, pp. 1567–1576, June 2002.
9. J. Kulik, W. Heinzelman, and H. Balakrishnan, "Negotiation-based protocols for disseminating information in wireless sensor networks," *Wirel. Netw.*, vol. 8, no. 2/3, pp. 169–185, 2002.

10. B. Horling, R. Vincent, R. Mailler, J. Shen, R. Becker, K. Rawlins, and V. Lesser, "Distributed sensor network for real time tracking," in *International conference on Autonomous agents*, (New York, NY, USA), pp. 417–424, ACM Press, 2001.

11. F. Zhao, J. Shin, and J. Reich, "Information-driven dynamic sensor collaboration for target tracking," *IEEE Signal Processing Magazine*, vol. 19, pp. 61–72, March 2002.

12. D. Li, K. Wong, Y. H. Hu, and A. Sayeed, "Detection, classification and tracking of targets in distributed sensor networks," *IEEE Signal Processing Magazine*, vol. 19, no. 2, pp. 17–29, 2002.

13. S. Pattem, S. Poduri, and B. Krishnamachari, "Energy-quality tradeoffs for target tracking in wireless sensor networks," in *International Symposium on Aerospace/Defense sensing Simulation and Controls, Aerosense*, April 2003.

14. A. Varga, "Omnet++: Objective modular network testbed in C++." http://www.omnetpp.org.

15. M. Ditzel and F. Elferink, "Low-power radar for wireless sensor networks," in *European Radar Conference*, September 2006.

A Zone-Based Clustering Method for Ubiquitous Robots Based on Wireless Sensor Networks

Kyungmi Kim[1], Hyunsook Kim[1], Young Choi[2], Sukgyu Lee[3], and Kijun Han[1]

[1] Department of Computer Engineering, Kyungpook National University
1370, Sangyuk-dong, Buk-gu, Daegu, 702-701, Korea
[2] Global Leadership School, Handong Global University
Bukgu, Pohang, 791-708, Korea
[3] Department of Electrical Engineering, Yeungnam University
214-1 Dae-dong Gyongsan, 712-749, Korea
kmkim@handong.ac.kr, hskim@netopia.knu.ac.kr,
ychoi@handong.ac.kr, sglee@yu.ac.kr,
kjhan@knu.ac.kr

Abstract. In this paper, we propose a topology configuration method for wireless sensor networks with an objective of well balancing energy consumption over all sensor nodes without generating any isolated sensor nodes. Our scheme has some attractive features: First, a high density node having good many neighbor nodes can be selected as a clusterhead in a zone. Second, reconfiguration of cluster can be carried out in a single zone, not all over network field, to reduce the number of nodes that participate in changing clusterheads. Third, multiple-hop transmissions between nodes or between clusterheads are possible. Finally, the proposed method can be applicable to various ubiquitous computing services including the ubiquitous robot (UR) based on wireless sensor networks. Simulation results show that our method outperforms LEACH and PEGASIS in terms of the system lifetime.

Keywords: wireless sensor network, ubiquitous robots, clustering, routing.

1 Introduction

Recently, the idea of wireless sensor networks has attracted a great deal of research attention due to the development of low cost and low power sensing devices with computational ability, wireless sensing and communication capabilities. They are especially useful in extremely hostile environments, such as near volcanically active sites, inside a dangerous chemical plant or in disaster area with a nuclear reactor. They also have advantages in inaccessible environments, such as difficult terrains, or on a spaceship[1]. Generally a sensor node consists of sensing elements, microprocessor, limited memory, battery, and low power radio transmitter and receiver. In wireless sensor networks, sensor nodes are usually unattended, resource-constrained, and unrechargeable. Since the batteries of the sensor nodes are not regularly rechargeable or not replaceable, the lifetime of a system is limited, and thus distributing power consumption to all nodes is a major design factor [2]. Therefore, locating sensor nodes over network fields efficiently is one of the most important

P. Havinga et al. (Eds.): EUROSSC 2006, LNCS 4272, pp. 25–38, 2006.

topics in sensor networks. Also, clustering approaches in wireless sensor networks have been proposed in to minimize the energy used to communicate data from nodes to the sink [3-5]. A good clustering scheme should preserve its structure of cluster as much as possible [6].

In the meantime, the limited power that the sensory nodes possess to collect data, transmit data to a mobile robot, raises the importance of implementing various power saving strategies when designing network architectures in a randomly distributed sensor network. The sensors distributed in the network are smart dust type miniature sensors expected to last for a longer duration of time. Therefore we need the mobile robot used to travel to collect data so that the power used for transmission by the sensor is optimized [7]. In most recent years, motivated by the emergence of ubiquitous computing technology as the next generation of computing paradigm, a new class of networked robot - ubiquitous robot (UR) - has been introduced. The URC (Ubiquitous Robot Companion) is a conceptual vision of UR which provides the services for person, anytime and anyplace in ubiquitous computing environment [8]. Since it is inherently based on ubiquitous environment with networked sensors and actuators, it can be considered as one of the most important emerging applications of sensor network. Figure 1 depicts the concept of URC with sensing, processing and acting abilities in wireless sensor networks to overcome the technical constraints and producing costs by utilizing sensors and remote computing server.

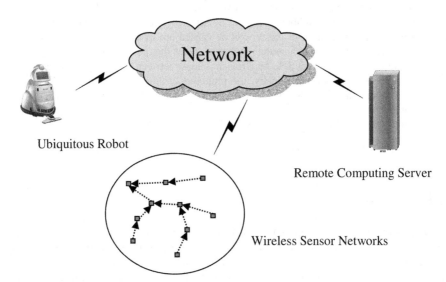

Fig. 1. The concept of Ubiquitous Robot Companion (URC)

As the technical challenges of wireless sensor networks for awareness of the environment are being solved, UR employs them instead of the on-board sensors to collect real-time information to facilitate its navigation which is crucial to many applications. Distributed networks which consist of sensors to perceive and send data measured in the surrounding environments are developed to reposition and organize sensors to acquire and deliver the corresponding information [9]. In fact, sensor networks can be successfully utilized for robot navigation [10-11]. The most important merits of sensor network in UR system include its ability to relay information from wide and distant regions of the environment with several distributed cheap sensors replacing expensive on-board sensors of UR. By using the network with external cheap sensors embedded in the environment replacing multiple robots equipped sensors, the context-awareness of the robot would be dramatically improved and lessen the burden of hardware cost [12]. Besides, remote computing server can be used as an external memory and processor of URC, which improves the robot intelligence and expands its applications and services.

Researchers are looking into merging robotics and wireless sensor network technologies. Therefore we are trying to join the UR into our clustering scheme. In this paper, we propose a scheme to evenly distribute clusterheads over the network field to reduce the energy consumption and the computational overhead. To distribute the clusterheads evenly, the network field is divided into several zones and the number of clusterheads to be included in each zone is determined in proportion to its area. Besides, we invite the UR as a device which aggregates the packets from all clusterheads and controls all the nodes. In our method, the sensed data is transmitted over multiple-hop path through clusterheads. Since reclustering is performed in a single zone independently, the computational overhead will be reduced as compared with the conventional approaches in which reclustering is carried out for all nodes in the network field every round.

This paper is composed as follows. We review some related work in section 2 and present an overview and discussion of our method in section 3. In section 4, we compare our method with the existing protocols and show the results. Finally, we conclude the paper in section 5.

2   Related Work

Heinzelman has proposed Low-Energy Adaptive Clustering Hierarchy (LEACH) for efficient routing of data in wireless sensor networks. In LEACH, the sensors elect themselves as clusterheads with some probability and broadcast their decisions. Each sensor node determines to which cluster it wants to belong by choosing the cluster-head that requires the minimum communication energy. The algorithm is run periodically, and the probability of becoming a clusterhead for each period is chosen to ensure that every node becomes a clusterhead at least once within $1/p$ rounds, where p is 5 percent of the number of all nodes [13].

The positive aspect of LEACH is the fact that the nodes will randomly deplete their power supply, and therefore they should randomly die throughout the network. On the other hand, the randomized cluster heads will make it very difficult to achieve optimal results. Since random numbers are utilized, the performance of the system will vary according to the random number generation and will not be as predictable as a system that is based on information that will lead it to make the best local decision [14].

A centralized version of LEACH, called LEACH-C, was proposed in [15]. Unlike LEACH, where nodes self-configure themselves into clusters, LEACH-C uses a centralized algorithm that employs the sink as a cluster formation controller. During the setup phase of LEACH-C, the sink receives information regarding the location and energy level of each node in the network. Using this information, the sink finds a predetermined number of clusterheads and configures the network into clusters. The cluster groupings are chosen to minimize the energy required for non-cluster-head nodes to transmit their data to their respective clusterheads as shown in Figure 2 (a) [15].

Power Efficient Gathering in Sensor Information Systems (PEGASIS) [7] enhances network lifetime by increasing local collaboration among sensor nodes. In PEGASIS, sensor nodes are arranged in a chain topology using a greedy algorithm so that each node transmits to and receives from only one of its neighbors. Every round, a randomly chosen node from the chain will transmit the aggregated data to the sink, thus reducing the per round energy consumption compared to LEACH [6]. PEGASIS is an interesting approach; however, there is the potential to achieve better performance for many applications because of three reasons: (1) the clustering is based on random cluster heads, (2) 100 % aggregation is not realistic for many applications, and (3) the chain described in PEGASIS is not an optimal routing mechanism in terms of the distance the data needs to traverse [6].

The core ideas of Base station Controlled Dynamic Clustering Protocol (BCDCP) [6] are the formation of balanced clusters where each clusterhead leads an nearly same number of sensor nodes to avoid clusterheads overload, uniform placement of clusterheads throughout the whole sensor fields, and utilization of cluster-head-to-cluster-head routing to transfer the data to the base station as shown in Figure 2 (b).

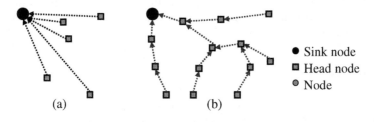

(a)   (b)

● Sink node
◨ Head node
◎ Node

(a) single hop (b) multiple hops

Fig. 2. Routing methods

Figure 3 shows typical examples of wireless sensor network in robot navigation. In Figure 3 (a), a robot navigates based on the information from sink node which possesses the global information on the environment. Since most of the wireless sensor networks with unrechargeable battery are considered disposable with the rate of energy consumption determining their operational lifetime, many alternatives for conserving power of each node have been considered.

Figure 3 (b) depicts a robot navigation using the information from the activated sensors, where corresponding nodes are activated only on request of the navigating robot. The robots may extend sensor networks to bring new activating sensors and move across the sensor field for reliable data collection, while the remaining sensors stay in idle mode to save energy. In addition, activating sensor nodes in Figure 3 (b) act as signposts for the robot to follow without a map or localization on the part of the robot. Since the robot interacts with nodes in the network locally to make navigation decisions based on which node it is near, context aware development of sensors is very important.

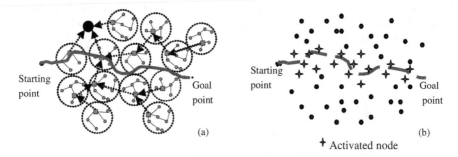

Fig. 3. Typical examples of sensor network in a mobile robot navigation

3 The Proposed Scheme

We propose a topology configuring scheme for wireless sensor networks with an objective of balancing energy consumption over all nodes in the network field without generating any isolated nodes. We call it 'zone-based clustering method for ubiquitous robots'. Our scheme starts from dividing a network field into several zones depending on the distance from the origin point. Each sensor node transmits data to the nearest neighbor node or clusterhead in each zone. Each clusterhead aggregates data and sends it to the nearest clusterhead in the neighboring zone towards the UR. The UR accumulates all packets from clusterheads. Since the UR acts as a central intelligent device that aggregates data streams from the deployed sensor nodes, interprets the data, and provokes intelligent actions. It can be considered as a sink node. The UR is an essential component with complex computational abilities. On the other hand, the sensor nodes are considered to perform very simple and cost effective functions.

Our scheme has three primary goals:

- prolonging network lifetime by evenly distributing clusterhead over the network,
- balancing energy consumption by selecting clusterhead in proportion to the area of each zone, and
- saving communication energy with multiple-hop

Several assumptions needed in our scheme are:

(1) All nodes in the network are uniformly distributed and quasi-stationary, (2) all nodes are homogeneous, energy constrained and location-aware, (3) all nodes are sensing at a fixed rate and always have data to send, (4) the UR controls clusterhead selection, and (5) all data sent by the previous nodes are aggregated by a constant bit size.

Our method organizes into three stages: (1) zone configuration stage, (2) clustering stage, (3) reclustering and data communication stage as shown in Figure 4. Zone configuration stage is executed just once at the time of network initialization to divide overall network into several zones. Clustering stage is also carried out one time to choose the initial clusterhead in all zones when the network is first deployed. Reclustering and data communication stage is performed for every round in a single zone independently to reduce the computational overhead as much as possible.

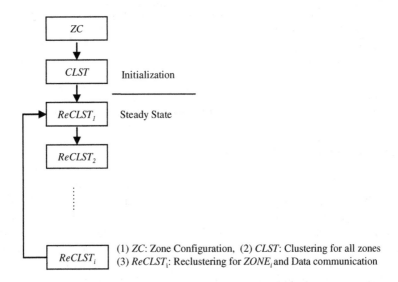

(1) *ZC*: Zone Configuration, (2) *CLST*: Clustering for all zones
(3) *ReCLST*$_i$: Reclustering for *ZONE*$_i$ and Data communication

Fig. 4. Procedure for our method

3.1 Zone Configuration Stage

The main operation in this stage is to divide the network field into several zones as shown in Figure 6. It is divided into several zones based on the zone range (*r*) which is determined by considering the network size, transmission range, and distribution density of the nodes. The first zone, denoted by *ZONE*$_0$, contains the UR and sensor nodes whose distances to the origin point are less than the zone range (r). The next zone, *ZONE*$_1$, contains sensor nodes whose distances to the origin point are greater than r but less than 2*r*. So, the i-th zone, *ZONE*$_i$, includes sensor nodes whose distances to the origin point are greater than $i \times r$ but less than $(i+1) \times r$. The last zone covers all remaining sensor nodes beyond the boundary of the previous zone. Thus, the total number of zones configured in the network is given by $(NETWORK\_RANGE)/r + 1$.

After zone configuration, the UR broadcasts the zone information to have each node know which zone itself is assigned to. The number of clusterheads in each zone is determined in proportion to the area of each zone. In *ZONE*$_0$, no clusterhead is

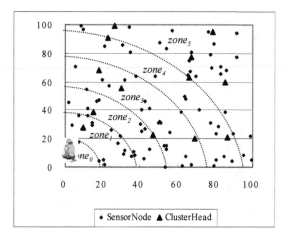

Fig. 5. Zone configuration

allowed because this zone has the UR. In $ZONE_1$, we can assign only one clusterhead. The number of clusterheads in $ZONE_2$ is proportional to the area of $ZONE_2$. For example, if the area of $ZONE_2$ is about two times than the area of $ZONE_1$, $ZONE_2$ has two clusterheads. The number of clusterheads in each zone is obtained by

$$N\_CH_i = \frac{Area(ZONE_i)}{Area(ZONE_1)}, \quad i \ge 2 \tag{1}$$

where N_CH_i is the number of clusterheads elected in $ZONE_i$.

3.2 Clustering Stage

This stage consists of the clusterhead(CH) selection, the cluster setup, and the formation of routing paths. The UR chooses clusterheads until the desired number of clusterheads in each zone is attained. We select a high density node, which has a good many neighbor nodes, as a clusterhead in each zone. Reclustering is performed for a single zone every round. After the number of rounds equal to a multiple of the number of zones, all clusterheads are replaced once for all zones. Cluster setup operation in this stage means that each node joins in the close clusterhead in the same zone. Once the clusters and the clusterheads have been identified, the UR determines the routing path for any two adjacent clusterheads as illustrated in Figure 6. All sensor nodes transmit data to the close neighbor node until reaching the clusterhead in the cluster.

3.3 Reclustering and Data Communication Stage

The primary tasks in this stage are reclustering for a single zone, data gathering, data fusion, and data forwarding. Reclustering is performed for a single zone every round. After the number of rounds equal to a multiple of the number of zones, all

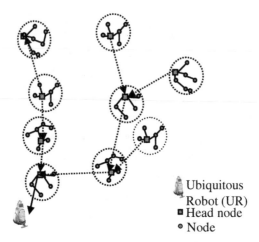

Fig. 6. Multi-hop between CH-to-CH and node-to-node

clusterheads are replaced once for all zones. All sensor nodes transmit the sensed information to their clusterhead by multiple-hop paths as shown in Figure 6. Once a clusterhead receives data from any nodes, it performs data fusion on the collected data to reduce the amount of raw data that needs to be sent to the UR. A sensor node transmits its data to the nearest neighbor node within the cluster that it belongs to. The neighbor node aggregates the data with its own data, and transmits it to the next node until reaching to cluster head. Similarly, the cluster head sends its aggregated data to the nearest cluster head in the next zone until arriving at the UR. There is an exception in case that all nodes in $ZONE_0$ where data is transmitted to the UR directly.

The data delivery model of wireless sensor networks can be classified several types: (1) continuous, (2) event-driven, (3) observer-initiated, and (4) hybrid. Our scheme is based on the continuous model. It means that all sensor nodes are assumed to carry out sensing operation at a fixed rate and always have data to send when they receive query messages from the UR. We assume that the UR request or the query occurs every round.

3.4   Extension of the Proposed Method

We can extend the proposed method to applications with a set of senor networks with one supervisory ubiquitous robot(SUR) with limited mobility in the global system and one sink node(UR) in each sensor network as shown in Figure 7, where the SUR has enhanced communication and computation capability over the UR in each sensor network, the performance of the global system is highly enhanced. In addition the level of autonomy exercised by URs in multi-UR system is increased.

In the future, we intend to more deeply explore the contextual configuration of sensor network by applying multiple sensor networks with limited mobility as shown

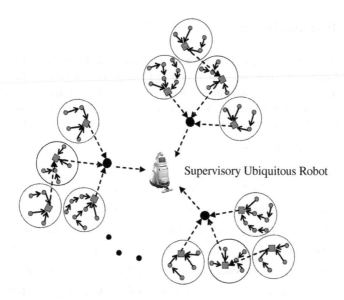

Fig. 7. The extended system with a Supervisory Ubiquitous Robot

in Figure 8. If we employ a set of mobile sensors in the network field, we can have more flexibility and efficiency to carry out clustering of sensors as well as more fault tolerance capability to compensate malfunctioning sensor nodes. Thus, it will enhance the feasibility and fault tolerance of the URC applications based on wireless sensor networks.

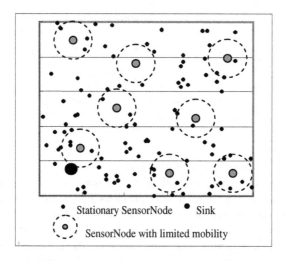

Fig. 8. An extended sensor network with a set of sensor.Nodes with limited mobility

4 Simulation and Results

To estimate the performance of our method, we compared its performance with other cluster-based protocols such as LEACH and PEGASIS. We simulated LEACH with a probability of 5% that each node elects itself clusterhead. As a radio model, we use the same one discussed in [14]. The energy costs for the transfer of k-bits data message between two nodes separated by a distance of r meters is given by,

$$E_T(k,r) = E_{Tx} \times k + E_{amp}(r) \times k \tag{2}$$

$$E_R(k) = E_{Rx} \times k \tag{3}$$

where $E_T(k,r)$ indicates the total energy for transmission of the source sensor node, and $E_R(k) = E_{Rx} \times k$ expresses the energy cost incurred in the receiver of the destination sensor node. The parameters E_{Tx}, E_{Rx} are the energy consumption for communication. $E_{amp}(r)$ is the energy required by the transmit amplifier to maintain an acceptable signal-to-noise ratio in order to transfer data messages safely. Also the energy cost for data aggregation is the set as $E_{DA} = 5nJ / bit / message$ [6].

Throughout the simulation, we consider a 100 x 100 network configuration with 100 nodes where each node is assigned an initial energy of 1.0J, the amount of transmission energy is 50 nJ/bit, transmit amplifier energy (E_{amp}) is 100 pJ/bit. The zone range is set by 19 (see Table 1).

Table 1. Simulation Parameters

Parameter	Value	Parameter	Value
Network size	100 x 100	Transmission energy	50 nJ/bit
Number of nodes	100	Data Aggregation energy	5 nJ/bit/message
Packet size	2000 bits	Transmit amplifier energy	100 pJ/bit/m2
Initial energy of a node	1 J	Zone range (r)	19

In simulations, all nodes are assumed to carry out sensing operation at a fixed rate and always have data to send when they receive query messages from the UR. It is also assumed that all data sent by the previous nodes are aggregated into a data segment with a constant size of 2000 bits. We assume that every node performs data aggregation when forwarding data to the next hop. So, once a node receives data from any sensor nodes, it performs data aggregation on the collected data to reduce the amount of raw data. Table 2 shows the number of clusterheads in each zone in the simulation with the network size of 100 m x 100 m.

Table 2. The number of clusterheads in each zone

Zone number	Zone area	Number of clusterheads (N_CH_i)	Number of clusterheads used in simulations
$ZONE_0$	283.39	0	0
$ZONE_1$	850.16	1.00	1
$ZONE_2$	1416.93	1.67	2
$ZONE_3$	1983.70	2.33	3
$ZONE_4$	2550.47	3.00	3
$ZONE_5$	2915.28	3.43	4

Figure 9 shows the average transmission distance from all nodes to the UR. From this figure, we confirm that our scheme produces a shorter transmission distance than those of LEACH and PEGASIS.

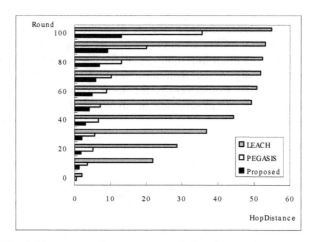

Fig. 9. The average distance to transmit data from nodes to the UR

This is because our method adopts a multi-hop routing path between any sensor node and the UR. Once the clusters and the clusterheads have been identified, the UR node chooses the routing path for any two adjacent clusterheads as illustrated in Figure 6. All sensor nodes transmit data to the close neighbor node until reaching the clusterhead in the cluster as depicted in Figure 6.

Figure 10 shows the number of rounds when a sensor node is dead for the first time and all sensor nodes are dead. The x-axis represents the number of rounds until the first or the last sensor node dies. This plot clearly shows that our scheme has more desirable energy expenditure than those of LEACH and PEGASIS. Also, we can see that our scheme offers a longer number of rounds to the first sensor node death. Also, our method outperforms LEACH and PEGASIS in terms of the system lifetime. As shown in Figure 6, the short transmission distance contributes to extending the number of rounds until the first sensor node and the last sensor node is dead.

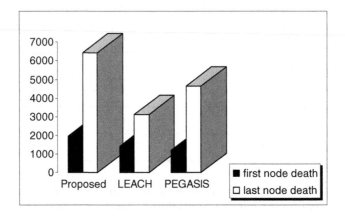

Fig. 10. The number of rounds until the first or the last sensor node dies

Figure 11 shows the amount of used energy at all sensor nodes for each round. This plot shows that our scheme offers an improvement as compared with LEACH and PEGASIS since we allow a multi-hop routing path for data transmission, the distance required for data transmitting are less than those of LEACH, and a node has sustained more rounds than LEACH. Consequently, the short transmission distance in our method makes a little energy consumption at all nodes. It lengthens the network lifetime.

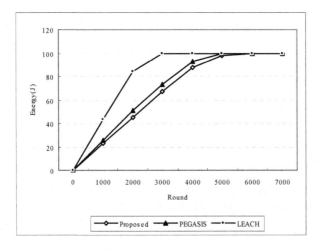

Fig. 11. The amount of used energy as time goes on

5 Conclusions

In this paper, we propose a method for selecting clusterheads, called 'zone-based clustering method for ubiquitous robots', in a wireless sensor network to balance the amount of energy consumption over all nodes without generating any isolated sensor nodes. The ubiquitous robot(UR) have the potential to greatly increase the feasibility of practical wireless sensor networks. In our method, the network field is first divided into several zones, and each zone includes clusterheads in proportion to its area, which contributes to distributing clusterheads evenly over the network. Simulation results show that our method outperforms LEACH and PEGASIS in terms of network lifetime. It is believed that our network configuration can be most efficiently utilized for the ubiquitous robot companion (URC) applications. In the future, we plan to extend our method to a more complicated and practical situation where multiple URCs are moving around the wireless sensor network.

References

1. Jamil Ibriq and Imad Mahgoub, "Cluster-Based Routing in Wireless Sensor Networks: Issues and Challenges," *In the Proceedings of the 2004 Symposium on Performance Evaluation of Computer Telecommunication Systems*, pp. 759-766, March 2004.
2. Pieter Beyens, Ann Nowe and Kris Steenhaut, "High-Density Wireless Sensor Network: a New Clustering Approach for Prediction-based Monitoring," *EWSN 2005 - Second European Workshop on Wireless Sensor Networks*, Istanbul, Turkey, 2005.
3. Siva D. Muruganathan, Daniel C.F. Ma, Rolly I. Nhasin, and Abraham O. Fapojuwo, "A Centralized Energy-Efficient Routing Protocol for Wireless Sensor Networks," *IEEE Communications Magazine*, vol. 43, pp. s8-13, March 2005.
4. Stephanic Lindsey and Cauligi S. Raghavendra, "PEGASIS: Power-Efficient Gathering in Sensor Information Systems," *Proceedings of the IEEE Aerospace Conference*, March 2002.
5. Gayathri Venkataraman, Sabu Emmanuel, Srikanthan Thambipillai, "DASCA : A Degree and Size based Clustering Approach for Wireless Sensor Networks," *IASTED International Conference on Networks and Communication Systems (NCS 2005)*, Thailand, April 2005.
6. Mannak Chatterjee, Sajal K.Das and Damla Turgut, "WCA: A Weighted Clustering Algorithm for Mobile Ad Hoc Networks," *Journal of Cluster Computing (Special Issue on Mobile Ad hoc Networks)*, 5(2): 193-204, April 2002.
7. Pathirana, P.N.; Black, T.J.; Nahavandi, S., "Path planning for sensor data collecting mobile robot," *Intelligent Sensors, Sensor Networks and Information Processing Conference, 2005. Proceedings of the 2005 International Conference on* 5-8 Dec. 2005 Page(s):313 – 317
8. Young-Guk Ha; Joo-Chan Sohn; Young-Jo Cho, "Service-oriented integration of networked robots with ubiquitous sensors and devices using the semantic Web services technology," *Intelligent Robots and Systems, 2005. (IROS 2005). Internatinal Conference on* 2-6 Aug. 2005 Page(s):3947 - 3952
9. Peter Corke, Ron Peterson, and Daniela Rus, "Coordinating Aerial Robots and Sensor Networks for Localization and Navigation," *Proceedings of the Seventh International Symposium on Distributed Autonomous Robotic Systems,* June, 2004.

10. Q. Li, M. De Rosa, and D. Rus, "Distributed algorithms for guiding navigation across a sensor network," *in proceedings of the 9th annual international conference on Mobile computing and networking*, pp.313-325, ASM Press, 2003.
11. K. K. O'Hara and T. J. Balch, "Distributed path planning for robots in dynamic environments using a pervasive embedded network," *in AAMAS*, pp. 1538-1539, July 2004.
12. H.Kim, Y.-J.Cho and S.-R.Oh, "CAMUS: A middleware supporting context-aware services foe network-based robots," *In Proceeding of the IEEE Workshop on Advanced Robotics and its Social Impacts(ARSO)*, 2005.
13. Wendi Rabiner Heinzelman, Anantha Chandrakasa, and Hari Balakrishnan, "Energy-Efficient Communication Protocol for Wireless Microsensor Networks," *Proceedings of the 33rd International Conference on System Sciences (HICSS '00)*, January 2000.
14. Mohamad Younis, Meenakshi Bangad and Kemal Akkaya, "Base-Station Repositioning for Optimized Performance of Sensor Networks," *Vehicular Technology Conference, IEEE 58th Volume 1, Issue,* 6-9 October 2003.
15. W. B. Heinzelman, A. P. Chandrakasan, and H. Balakrishnan, "An Application-Specific Protocol Architecture for Wireless Microsensor Networks," *IEEE Transactions on Wireless Communication, vol. 1, no. 4*, pp. 660-670, October 2002.

A Simulation Study of Integrated Service Discovery

Gertjan P. Halkes, Aline Baggio, and Koen G. Langendoen

Faculty of Electrical Engineering, Mathematics and Computer Science
Delft University of Technology, The Netherlands
{g.p.halkes, a.baggio, k.g.langendoen}@tudelft.nl

Abstract. The research in the field of service discovery in mobile ad-hoc networks is characterised by a lack of quantitative research. Many ideas have been put forward but few have been tested, either in simulation or real life. This paper fills part of that void, by comparing through simulation a simple broadcast-flood protocol, an integrated routing and service-discovery approach, and a global-knowledge based approach. The results show that using an integrated approach can achieve a similar level of performance as a global-knowledge based approach.

Keywords: Service discovery, Mobile Ad-Hoc Networks, Simulation.

1   Introduction

The field of service discovery is gaining more and more attention in the mobile ad-hoc network (MANET) research community [1,2,3,4,5]. However, the field is characterised by a lack of quantitative research. Many ideas have been put forward but few have been tested, either in simulation or real life. This paper fills part of that void, by comparing three different schemes through simulation: a simple broadcast-flood protocol, an integrated routing and service-discovery approach, and a global-knowledge based approach.

The simple flood protocol is an unoptimised service-discovery protocol. It is important as it is a natural extension of the Service Location Protocol (SLP) [6], which was developed for fixed-infrastructure networks, into the MANET domain.

Integration of routing and service-discovery is an idea that has been forward by Koodli et al. [5]. By performing service discovery in the same way as route discovery, nodes can accumulate routing information while performing service discovery. If a service provider wishes to reply to a received service request it does not have to perform a route discovery for the originator of the request, because it already has the required routing information. This is a big advantage over a purely application-layer based approach, like the simple flood protocol, where all service providers that wish to reply to a service request have to do route discovery.

The global-knowledge approach uses an oracle to determine which service providers are available in the network and to locate the service provider that is most suitable to communicate with, that is, the closest one. The oracle serves

P. Havinga et al. (Eds.): EUROSSC 2006, LNCS 4272, pp. 39–53, 2006.

as an upper bound on the performance of any (integrated) service discovery protocol; similarly, the simple flood protocol provides a lower bound. In our experimental evaluation we have studied the effect of node mobility, service request rates, node density, and lifetime of cached service entries.

The contribution of this paper is twofold: firstly, it provides a comparison through simulation of integrated service-discovery with an unoptimised service-discovery protocol and a global-knowledge approach. Secondly, it provides a benchmark for further comparisons.

The rest of this paper is organised as follows: Section 2 gives a short overview of the research efforts in service discovery in MANETs. It is followed by background information on routing in Section 3. Section 4 details the service discovery protocols used in our simulation study. Section 5 presents the design of our experiments. Section 6 shows the results from our experiments, analyses the major trends, and discusses their significance. Finally, Section 7 lists our conclusions.

2   Related Work

Service discovery in fixed-infrastructure networks has received quite some attention. Standards are now being developed, the most important of which is the Service Location Protocol (SLP) [6]. The SLP protocol has two modes of operation: centralised and distributed. The centralised mode uses one or more service directories. Service providers register their services with the service directories. If a client wants to discover a server it contacts the service directories and requests a list of matching servers.

In distributed mode, service directories are not used. To find a service, a client simply broadcasts a service query on the network. If a service provider receives such a query, it sends a unicast message to the originator. Optionally, one can use multicast instead of broadcast, so as to limit the network traffic.

The centralised mode of SLP does not match with the ad-hoc nature of MANETs. Using the distributed mode, however, is feasible. Reusing an implementation of SLP meant for fixed-infrastructure networks is possible, by replacing the local broadcast with a broadcast flood. The resulting protocol is similar to the Nom [3] protocol. The main difference is that the Nom protocol also implements a cache of previously-seen service bindings. This service cache reduces the number of service requests sent to the network by allowing the reuse of previously gathered information.

Many papers have been written on the field of service discovery in MANETs. We now present the most important proposals.

The Intentional Naming System (INS) [1] is one of the first proposals for service discovery in multi-hop ad-hoc networks. INS integrates routing and service discovery, but does so using an overlay network. To create the overlay network, a central component is used. The overlay network is used both as a replicated distributed service directory, and as a network of forwarders. INS is a proactive protocol in the sense that services are advertised to and stored in a service directory, before the service information is requested. INS has a number of drawbacks.

First it has a single point of failure in the form of the central component used to build the overlay network. Second, keeping the service directory up to date, even when it is not used, can incur significant network traffic.

Another early proposal on how to implement service discovery in mobile ad-hoc networks has been put forward by Koodli et al. in a now expired draft RFC [5]. The authors propose to integrate service discovery and routing. By doing so, one leverages the existing experience with routing protocols to create an efficient service discovery protocol.

Several other protocols have been suggested that integrate a limited form of routing in the protocol itself, such as (GSD [2], CARD [4]. GSD reuses many of the ideas of the AODV protocol [7], while CARD implements large parts of the TRANSFER [8] routing protocol. However, these protocols are in essence duplicating some of the work of the routing layer which is inefficient. Moreover, the routing information gathered by the service discovery protocol cannot benefit other traffic in the network.

In a follow-up paper the authors of GSD [2] extend the integration with routing to also include subsequent communication with the service provider [9]. As an extra feature, the authors propose automatic re-routing to another available service provider if the route to the selected service provider breaks. Although this approach uses the routing information gathered during service discovery in subsequent communications with the service provider, other types of traffic still cannot benefit from this routing information. Moreover, the re-routing of traffic to another service provider only works when there are multiple providers delivering an indistinguishable service.

To demonstrate the advantages of integrated service discovery, a thorough study of the proposed protocols is needed, nevertheless performance has received only limited attention. Varshavsky et al. [10] have done a worst-case packet-count analysis and an experimental (in simulation) comparison with variants of the Service Location Protocol (SLP) [6]. In a later paper [11], Varshavsky et al. evaluate service-selection mechanisms, but only compare with centralised SLP variants. A paper by Garcia-Macias et al. [12] provides a very limited case study of integration with AODV vs. the Nom protocol. Both studies conclude that using integrated service-discovery can significantly reduce the number of messages needed for service discovery. This is also confirmed by the results reported in this paper, which includes a set of simulation experiments covering a wide range of parameters, for example, node density and speed.

3   Routing Background

As we will be presenting service-discovery protocols based on routing protocols, we now present some background information. Readers familiar with basic routing protocols like AODV [7] and DSR [13] may proceed with Section 4 immediately.

Routing protocols can be categorised into reactive and proactive routing protocols. Reactive routing protocols do not maintain routing information for the entire network, but only start communicating when a route is required.

Proactive routing protocols do maintain routing information for the entire network, and therefore don't have to communicate to find a route at the time the route is needed. On the other hand, maintaining this route information means that proactive routing protocols need to communicate constantly, even if no routes are needed.

The most well-developed reactive routing-protocol for MANETs at this time is the Ad-hoc On-demand Distance Vector (AODV) protocol. Nodes using AODV maintain a routing table. This routing table contains a next-hop address for all nodes to which a route is known. When a node using the AODV routing protocol needs to know a route to another node in the network, it first checks its routing table. If an entry is present, the message is forwarded to the node mentioned in the routing entry. Otherwise, a route discovery procedure is initiated.

A node performs a route discovery by broadcasting a Route Request (RREQ) message to its one-hop neighbours. Sending such a message is achieved by sending an RREQ message with a Time-To-Live (TTL) of one. If one of them has routing information for the requested address, it replies with a Route Reply (RREP) message. If, after a timeout, no neighbour has replied with an RREP the RREQ message is resent, this time with an increased TTL. On receiving a message with a TTL greater than one, a node rebroadcasts the message if it can not supply the originator with the desired routing information. A node records the sender of the message as the next hop for sending messages to the originator of the RREQ. This way a so-called reverse route is set up. This reverse route can then be used to send the RREP message. As long as the originator does not receive an RREP message within the timeout period for the TTL set in the RREQ message, it increases the TTL in the RREQ up to a certain maximum. This technique is called expanding-ring search.

Another prominent reactive routing-protocol is the Dynamic Source Routing (DSR) protocol. In contrast to the AODV protocol, the routing table of the DSR protocol contains route information for entire routes and not just the next-hop address for the different destinations. This can be implemented in several ways. Originally, the authors proposed to use a so-called route cache, whereby the routing table contained full routes to each known destination. This was later deemed impractical for larger networks, so the link cache was proposed. In this scheme, a node stores which links are available between nodes. When a route is needed, the link information is used to build a complete route to the destination.

The DSR protocol uses source routes in its messages. This means that each message contains the untraveled part of the route to the destination. Like AODV, the DSR protocol uses RREQ messages to gather route information. The DSR RREQ messages contain the route back to the originator so that the receiver of the message can also send a message back to the originator. By default the DSR protocol does not use the expanding-ring search. Instead a node starts with asking its one-hop neighbours first, and if they don't reply within a preset timeout the node sends an RREQ message with the TTL set to a predefined maximum. The DSR protocol uses overhearing (aka. promiscuous mode) to allow nodes that are near but not on a path to gather route information. Note that the use of source routes is beneficial in this situation.

4   Service-Discovery Protocols

For our performance study, we have chosen two reactive routing protocols, namely AODV and DSR. These represent the most well-developed protocols currently available. Their operation is sufficiently different to warrant separate treatment. We have implemented integrated service discovery for both protocols and use the prefix SD to distinguish the integrated service discovery protocol from the original routing protocol.

As for comparison, we have chosen two extremes. A simple broadcast-flood protocol, which is the natural extension of distributed SLP into the MANET domain and also resembles the Nom protocol. This represents completely unoptimised service discovery. The other extreme is represented by a global-knowledge approach where each node knows all available services and the physical distances to each of these services. Communication is kept as local as possible, and service discovery is essentially a non-operation.

Our simulations do not include any proactive routing protocols as the effect of integration can be easily estimated analytically. Service information is not as volatile as routing information, therefore service information updates can be sent much less frequently than routing information updates. This means that the impact of disseminating service information in a manner similar to the dissemination of routing information gives a small and constant overhead.

In the following sections we will give more detail on each of the service-discovery protocols included in our simulations.

4.1   SD-AODV and SD-DSR

The SD-AODV and SD-DSR protocols have been implemented in the spirit of the routing protocols they extend. Two extra message types have been introduced, i.e. Service Request (SREQ) and Service Reply (SREP). The difference between an SREQ and an RREQ is that the target specified in the message is not an address, but a service description. An SREP differs from an RREP in that it also includes a service description of the offered service.

The forwarding and handling of SREQ and SREP messages is implemented like the forwarding and handling of RREQ and RREP messages in the original protocols. For example, in SD-AODV, SREQ messages use the same expanding-ring search technique used for RREQ messages. As the AODV protocol does not use overhearing, neither does SD-AODV. Conversely, DSR does use overhearing, therefore so does SD-DSR. Note that the SREQ and SREP messages also create entries in the routing tables in the same way as RREQ and RREP messages do.

The main difference between the handling of service related messages vs. route related messages is in the dissemination of SREP messages by so-called intermediate nodes. An intermediate node is a node that receives an SREQ message, but is neither the source nor the target of the SREQ message. SD-AODV and SD-DSR impose an additional constraint on the dissemination of SREP messages by an intermediate node: both a valid service description must be cached and a valid route to the target must be available. In SD-DSR only a one-hop neighbour of the source of the SREQ message may issue an intermediate-node reply. Experiments showed that allowing all intermediate nodes to issue replies increases the total number of messages sent, instead of decreasing it.

In both protocols, the handling of the service information differs slightly from the handling of the routing information. Each node has a service cache for storing service bindings. If an SREP message is received, the service description from the SREP message is used to create or update a service cache entry. The lifetime of the service-cache entry is determined from the received message. If the service cache entry already exists, the maximum of the lifetime of the existing entry and the lifetime in the message is taken as the new lifetime. Otherwise, the lifetime is copied form the received message.

The service cache is used to check for known service bindings, before initiating an SREQ message. If the service cache contains a valid and matching service description, no SREQ message is sent and the cached binding is returned to the application. However, should the application find that none of the bindings retrieved from the service cache could be used to contact a server after repeated attempts, the node requests a true service discovery to be initiated by the service-discovery protocol.

4.2   Flooding

The flooding protocol is the simplest of the service-discovery protocols. It uses the same service-caching regime as the SD-AODV and SD-DSR protocols. When a service query needs to be injected into the network, it simply initiates a network-wide broadcast flood. Intermediate nodes only pass on the request, even if they do have a valid service binding in their cache. When a request reaches a server that offers a matching service, this server sends a unicast message back to the source of the request.

As an optimisation, intermediate nodes that forward a service reply can inspect messages and cache the service binding. This ensures that a future request generated at the intermediate node can be satisfied by inspecting the cache, thus preventing a broadcast flood.

The flooding protocol is used in combination with both the AODV and the DSR routing protocol. When using the flooding protocol in combination with the DSR routing protocol, overhearing becomes a realistic option. The DSR protocol already uses overhearing itself, therefore it would not cost extra energy to also allow other protocols access to the gathered information. We have implemented a variant of flooding that incorporates overhearing.

4.3   Global Knowledge

When using the global knowledge approach, all nodes know which services are available on which server. The minimum hop-count to all the servers is available to all nodes at all times as well. All this information is provided by an oracle. By using this information, a node can select the closest server that provides a desired service without performing any communication. However, the oracle does not provide routing information. The global-knowledge approach is therefore used in combination with both the AODV and the DSR routing protocol. By not providing routing information, we obtain a means to measure the service discovery overhead of the other protocols.

5   Simulation Design

Our experiments were conducted using the QualNet wireless network simulator [14]. Each simulation has been run 10 times with different random seeds, and simulates 1000 seconds. We used the 802.11 MAC and physical layer, with a radio range of 140 meters. The QualNet simulator includes models for the AODV and DSR protocols. Unfortunately the default AODV model contained a bug, whereby an expired route entry would be revived on reception of a packet from the node named in the routing entry. Our experiments use a fixed version of the AODV model. The default DSR model uses a path cache. We have replaced this with a link cache because of its smaller memory footprint.

For the basic stationary network experiments, we used an area of 1000×1000 meters. For the experiments with mobile networks, we increased the area size to 1200×1200 to counter the centring effect of the Random Waypoint mobility model. All our experiments involve 100 nodes, 50 of which are clients. The number of servers is three, except for the first experiment, where there is only one server.

When mobility is used, we use the Random Waypoint model [13]. We set minimum speed V_{min} and maximum speed V_{max} to 1 and 5 m/s respectively, and set the maximum pause time to 30 seconds. V_{min} is not set to 0 to avoid the nodes in the network slowing down to the point where it becomes almost a stationary network [15].

Client nodes repeatedly request and use a service provided in the network; a client is modelled as a parametrised Poisson process specifying the service request rate. After a service is discovered by a client, it chooses one of the servers it has heard of and sends it a unicast message. To select the nearest server, the client sorts the servers by hop count using information from the routing layer, or in the case of the global-knowledge approach by distance. The server responds, also with a unicast message. This step models the communication between a client and a server for which the service discovery was initiated. If this communication fails even after two retries at the application level, the client selects another server from its list. If there are no more entries in the server list and the client has so far used only cached results, it asks the service discovery protocol to update its cache by issuing a new service request. If, after trying the servers this last step yields, communication with a server has still not succeeded, the client gives up.

As described in Section 4 all the service discovery protocols simulated use a service cache. The lifetime of entries in the service cache is set to 120 seconds.

This setting strikes a balance between saving gathered information on the one hand, and the volatility of the network on the other hand. In one of our experiments, we investigate the effect of different values of the service cache lifetime (see Section 6.4).

6   Results

This Section presents the results of our experiments. We start with a simple stationary network with a single server, and gradually explore more complex situations. First, we increase the number of servers. Then, we add mobility (Section 6.2) and explore the effect of low node density (Section 6.3). Finally, in Section 6.4, we show the effects of varying the lifetime of entries in the service cache.

6.1   Stationary Network

For our first experiment, we start with a stationary network, with a single randomly located server. Figure 1 shows the absolute number of packets sent in the network. The protocols using AODV clearly require more messages to find service providers and communicate with them than the protocols running on DSR. This is a characteristic of the routing protocols themselves and is not specific to service discovery [16].

Figure 2 shows the number of packets sent for the protocols using AODV, normalised by the global-knowledge approach. It is clear that the integrated SD-AODV protocol has similar performance as the global-knowledge approach. In fact, it sometimes performs slightly better. This is due to minor implementation differences in the sending of SREQ and SREP messages with respect to the RREQ and RREP messages.

The flood protocol performs worse than either the global-knowledge approach or the integrated SD-AODV protocol. There are two reasons for this. Consider a node A that is close to a node B. When node A has performed a successful service discovery, it has service and routing information for reaching the server. If node B is not on the route from A to the server, it will not have any information for reaching the server. In the case of SD-AODV or the global-knowledge approach, node B is able to gather the required information using the expanding-ring search technique. As node B is close to node A, this requires only a few packets. However, in the case of the flood protocol node B would initiate a full network flood, which needs many packets.

The second reason is that once the flood initiated by node B reaches the server, and the server wishes to send a response, it does not have any routing information for node B. This means the server needs to initiate an RREQ, which again uses many packets. Avoiding these extra RREQs is the most important reason for implementing integration with the routing protocol.

For higher request rates, the number of SREQs initiated by the flood protocol on AODV does not increase because of the use of caching. The total number of messages used for communication does increase as the clients communicate with

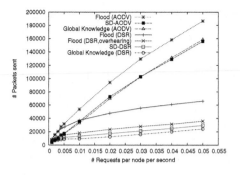

Fig. 1. Number of packets sent in a static network with one server

Fig. 2. Number of packets sent in a static network with one server for all protocols using AODV for routing, normalised by the global knowledge approach

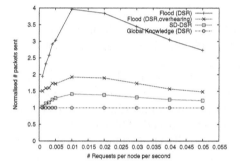

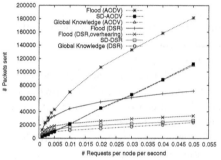

Fig. 3. Number of packets sent in a static network with one server for all protocols using DSR for routing, normalised by the global knowledge approach

Fig. 4. Number of packets sent in a static network with three servers

the server more often. The net effect is that the flood protocols performs better with respect to the global-knowledge approach for higher request rates.

Figure 3 shows the number of packets sent for the protocols using DSR, again normalised by the global-knowledge approach. Note however, that this is now the global-knowledge approach on top of the DSR protocol and not on top of the AODV protocol as in Figure 2. In this case we see that the integrated SD-DSR protocol cannot reach the same level of performance as the global-knowledge approach. This is explained by the difference in handling and using routing and service information, in combination with the overhearing used in (SD-)DSR. As far as RREQs and SREQs are concerned, the handling is mostly similar. However, when a node that has overheard routing information subsequently uses that information to send a packet, it thereby disseminates routing information to its neighbours. However, when using service information to communicate with a service provider the node does not disseminate this service information. This

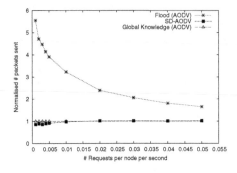 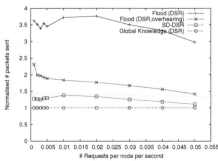

Fig. 5. Number of packets sent in a static network with three servers for all protocols using AODV for routing, normalised by the global knowledge approach

Fig. 6. Number of packets sent in a static network with three servers for all protocols using DSR for routing, normalised by the global knowledge approach

means that in some cases the global-knowledge approach has all the information it needs (i.e., the routing information) while SD-DSR still requires an SREQ to discover services.

At the lowest simulated request rate, the SD-DSR protocol achieves the same performance as the global-knowledge approach. This is because the request rate is so low that any route information spread by a communication between a client and a server times out before the next request is issued. Therefore, the clients will always have to initiate an RREQ or SREQ (SD-DSR) which uses the same number of packets.

Clearly the flood protocol is at a disadvantage when it does not use overhearing. However, even when using overhearing, it still suffers from the same problems as described for the combination of the flood protocol with the AODV routing protocol. Of course the exact effect is different as DSR does not use expanding-ring search, but a simple two-stage search. The better relative performance at very low request rates is again due to the route information timing out before a new request arrives.

Figures 4 through 6 show the results of simulations in the same static network, but now with three randomly located servers all providing the same service. In Figure 5, we can see that for the case with few service requests SD-AODV performs better than the global-knowledge approach. The fact that the global-knowledge approach chooses the closest server to communicate with is actually a slight disadvantage here. Where SD-AODV finds service and routing information that was spread by a previous service request of a nearby node, the global-knowledge approach is forced to find route information to the closest server. This routing information may not be found as nearby as the information that SD-AODV uses. Although SD-AODV communicates with a server further away, it uses fewer packets to find it. If communication were to continue between client and server, the shorter route that the global-knowledge approach sets up would ensure that the balance tips in its favour.

The DSR-based protocols behave mostly the same as for the single server case. The only difference is that for very low request rates the SD-DSR protocol and the flood protocol now perform worse, with respect to the global-knowledge approach. This difference is caused by the fact that the global-knowledge approach requests a route to a single server, while the service request from both the flood protocol and SD-DSR targets all servers in the network. As both SD-DSR and the flood protocol use broadcast floods, they will reach all the servers in the network. All these servers then have to send a reply, which is the cause of the extra messages.

We have also conducted experiment with three servers providing three different services, and with three servers each providing the same set of three servers, both with and without mobility. The results are so similar to the results already provided we do not show them here.

6.2   Mobility

As we are considering *mobile* ad-hoc networks, we introduce mobility into our experiments. Figure 7 shows the results of simulations with all nodes moving according to the Random Waypoint model. The most striking feature of this graph is its similarity to Figures 1 and 4.

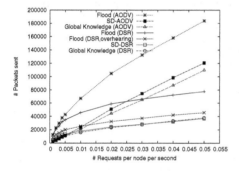
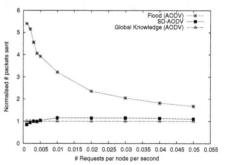

Fig. 7. Number of packets sent in a mobile network with three servers

Fig. 8. Number of packets sent in a mobile network with three servers for all protocols using AODV for routing, normalised by the global knowledge approach

If we further compare Figures 5 and 8 we can see that for protocols using AODV, mobility changes only one thing about the relative performance of the service discovery protocols: the SD-AODV protocol starts to perform worse. The relative performance decrease is indirectly caused by the expanding-ring search technique. The expanding-ring search technique has the characteristic that it stops searching once a single answer has been found. For routing this obviously is not a problem, as there should be only one node with a given address. However, in the case of service discovery, knowing about more than one server is beneficial.

When a node moves around, a different server from the one previously discovered may come (much) closer, and thereby becomes a much more attractive partner for communication. SD-AODV may not know about the closer server, and will try to keep communicating with the server it does know about. However, longer routes are also more prone to breaking and maintaining these longer routes therefore increases the number of packets sent. The longer routes themselves add to the number of packets sent, simply by requiring more packets to be sent to let one message travel between client and server.

In Figure 9, we see that all the service-discovery protocols on top of DSR perform better with respect to the global-knowledge approach than for the static network with three servers in Figure 6. In fact, SD-DSR performs just as well for higher request rates. The global-knowledge approach always chooses the nearest server to communicate with. However, the communication with the nearest server may be very unreliable for an amount of time because of an unreliable link on the route. Then, each time a node tries to communicate with the nearest server it has a high chance of failure. If after several retries it determines that the chosen server is unreachable, it tries to communicate with the second nearest server. All of this also holds for the SD-DSR protocol, but where the global-knowledge approach will always try to communicate with the nearest server, the SD-DSR protocol will remove an unreachable server from its cache and will not find the same problem over and over again. For low request rates this problem does not exist, because the unreliable links disappear before a new request is issued.

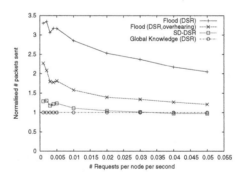

Fig. 9. Number of packets sent in a mobile network with three servers for all protocols using DSR for routing, normalised by the global knowledge approach

We have also conducted experiments with different settings of the pause time and maximum speed, and also with a different mobility model. The results showed the same effects as described above, so we omitted them for brevity.

6.3   Node Density

Our next experiment shows the effect of reducing the node density in the network. Figure 10 shows the result when the area size is increased from 1200×1200

to 1600×1600, thereby increasing the total area by a factor 1.8. The mobility induced problems are exacerbated by the decreased connectivity. In particular, the
persistence to use the closest server for each new request by the global-knowledge
protocol is causing a significant decrease in performance. Furthermore, the redundancy in routing is decreased, thereby increasing the number of route errors
and packets sent. The best example of the increased problems is the DSR-based
global-knowledge approach. It now performs worse than SD-DSR *and* the DSR-
based flood protocol with overhearing for high request rates.

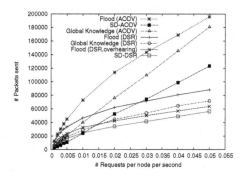

Fig. 10. Number of packets sent in a mobile network on a large area

6.4   Cache Lifetime

The final parameter we varied was the cache lifetime of the entries in the service
cache. We performed this experiment in the 1200×1200 network with mobility.
Figure 11 shows the results for the request rate of 0.01 requests per node per
second. The graphs for the other request rates show the same trends.

For all but the SD-AODV protocol, increasing the cache lifetime decreases the
number of packets sent. The flood and SD-DSR protocols know all the servers
in the network, after performing one service-discovery phase. Therefore, it is not

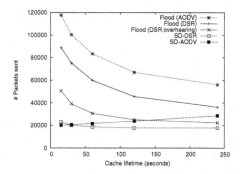

Fig. 11. Number of packets sent in a mobile network for different service-cache lifetimes
at request rate 0.01 requests per node per second

beneficial for these protocols to discard the service information. The more often these protocols discard the service information, the more often they have to do service discovery, at the expense of more packets.

The SD-AODV protocol has one disadvantage. As described in Section 6.2, the SD-AODV protocol mostly finds only a single server. In combination with mobility, this can cause a node to persist using a previously found server while another server is now nearer by. Doing service discovery more often therefore ensures that SD-AODV will regularly rebind to the closest server. Keeping communication local means that fewer packets need to be sent. The extra overhead of the service discovery is offset by the reduction of route breaks and the shorter routes. Note that the service discovery overhead is also limited, as SD-AODV uses the expanding-ring search technique.

7   Conclusions

In this paper, we presented extensive simulations comparing a simple flooding-based protocol, integrated service-discovery protocols and a global-knowledge approach for service discovery. We tested each of these protocols in combination with both the AODV and DSR routing protocols. The results show that the integration of service discovery with the routing protocol out-performs the simple flooding protocol for a large set of parameters.

The integration of service discovery with the AODV protocol can yield results on par with the global-knowledge approach. However, we also showed that to reach this performance with SD-AODV in mobility scenarios we needed to decrease the service-cache-entry lifetime, instead of increasing it. The cause of this counter-intuitive setting lies in SD-AODV's expanding-ring search. The expanding-ring search causes service requests to find only a limited number of services. In combination with mobility this causes nodes to make a suboptimal choice on which server to use once a node has moved. Frequent rediscovery mitigates this effect.

DSR uses source routing and overhearing. Because of this, sending a message to a server spreads route information to all nodes within communications range. Nevertheless, that message does not spread service information. This means SD-DSR is at a disadvantage compared with the global-knowledge approach. However, the persistence to use the nearest server can prove a problem for the global-knowledge approach when routing to that server is problematic. The result is that SD-DSR can perform both better and worse than the global-knowledge approach depending on the exact network conditions, but the performance never diverges greatly.

More generally, we conclude that the integrated service-discovery protocols are efficient and are therefore a much more suitable standard for comparing the efficiency of service-discovery protocols than the flooding protocol. The integrated protocols achieves a level of performance close to that of the global-knowledge approach. The only drawback is that for the integrated service-discovery protocols an implementation is required per routing protocol. Yet, the extensions to the routing protocol are sufficiently straightforward to allow rapid implementation.

References

1. Adjie-Winoto, W., Schwartz, E., Balakrishnan, H., Lilley, J.: The design and implementation of an intentional naming system. In: Proc. of the 17th ACM Symposium on Operating Systems Principles (SOSP'99). (1999) 186–201

2. Chakraborty, D., Joshi, A., Yesha, Y., Finin, T.: GSD: a novel group-based service discovery protocol for MANETS. In: Proc. of the 4th IEEE Conf. on Mobile and Wireless Communications Networks (MWCN 2002). (2002) 140–144

3. Doval, D., O'Mahony, D.: Nom: Resource location and discovery for ad hoc mobile networks. In: Proc. of the First Ann. Mediterranean Ad Hoc Networking Workshop (Med-Hoc-Net 2002). (2002)

4. Helmy, A., Garg, S., Pamu, P., Nahata, N.: Contact-based architecture for resource discovery (CARD) in large scale MANets. In: Proc. of the 17th Int. Parallel and Distributed Processing Symposium (IPDPS 2003). (2003)

5. Koodli, R., Perkins, C.E.: Service discovery in on-demand ad hoc networks. Internet draft, IETF Manet Working Group (2002) draft-koodli-manet-servicediscovery-00.txt (expired).

6. Guttman, E., Perkins, C.E., Veizades, J., Day, M.: Service location protocol, version 2. RFC 2608, IETF Network Working Group (1999)

7. Perkins, C.E., Belding-Royer, E.M., Das, S.: Ad hoc on-demand distance vector (AODV) routing. RFC 3561, IETF Network Working Group (2003)

8. Helmy, A.: TRANSFER: Transactions routing for ad-hoc networks with efficient energy. In: Proc. of the IEEE 2003 Global Communications Conference (GLOBECOM 2003). (2003) 398–404

9. Chakraborty, D., Joshi, A., Yesha, Y.: Integrating service discovery with routing and session management for ad hoc networks. Ad Hoc Networks, Elsevier Science (2004)

10. Varshavsky, A., Reid, B., de Lara, E.: The need for cross-layer service discovery in manets. Technical Report CSRG-492, Department of Computer Science, University of Toronto (2004)

11. Varshavsky, A., Reid, B., de Lara, E.: A cross-layer approach to service discovery and selection in MANETs. In: Proc. of the Second Int. Conf. on Mobile Ad-Hoc and Sensor Systems (MASS 2005). (2005)

12. Garcia-Macias, J.A., Torres, D.A.: Service discovery in mobile ad-hoc networks: better at the network layer? In: Proc. of the Int. Conf. Workshops on Parallel Processing (ICPP 2005). (2005) 452–457

13. Johnson, D.B., Maltz, D.A.: Dynamic source routing in ad hoc wireless networks. In Imielinski, Korth, eds.: Mobile Computing. Volume 353. Kluwer Academic Publishers (1996)

14. Scalable Network Technologies, Inc.: QualNet 3.8 user's guide (2004) http://www.scalable-networks.com/.

15. Yoon, J., Liu, M., Noble, B.: Random waypoint considered harmful. In: Proc. of the Twenty-Second Ann. Joint Conf. of the IEEE Computer and Communications Societies (INFOCOM 2003). Volume 2. (2003) 1312–1321

16. Broch, J., Maltz, D.A., Johnson, D.B., Hu, Y.C., Jetcheva, J.: A performance comparison of multi-hop wireless ad hoc network routing protocols. In: Proc. of the 4th Ann. ACM/IEEE Int. Conf. on Mobile Computing and Networking (MobiCom'98). (1998) 85–97

Context Dissemination and Aggregation for Ambient Networks: Jini Based Prototype

Kazimierz Bałos, Tomasz Szydło, Robert Szymacha, and Krzysztof Zieliński

Department of Computer Science
AGH University of Science and Technology
{kbalos, tszydlo, szymacha}@agh.edu.pl, kz@ics.agh.edu.pl

Abstract. This paper considers implementation of ContextWare architecture proposed for context dissemination and aggregation for Ambient Networks. ContextWare is a model for context processing dealing with sources of context localization and identification, acquiring context data from various sources, and its distribution to interested context consumers. Implementation requirements of this highly dynamic system with Jini has been proposed, evaluated and compared to other approach based on DHT. The extension of ContextWare with ontological description of context information has been also described. The presented concepts have been illustrated by the case study.

1 Introduction

Ambient Networks (AN) are aimed at enabling the co-operation of heterogeneous networks on demand, transparently to the potential users, and without the need for pre-configuration or offline negotiation between network operators [1]. Context-awareness facilitated by context information collecting, dissemination and management will play an essential role for AN activity. The general architecture of the system providing such functionality is *ContextWare* proposed by AN Project [1,23]. *ContextWare* could be considered as a middleware model for context processing dealing with sources of context localization and identification, acquiring context data from various sources, distribution context information to interested context consumers, caching and storing context in data base.

The existing specification of *ContextWare* is generic and does not rely on any particular implementation technology. It also does not investigate the inter-domain cooperation and interoperability problems [10] related to heterogeneity of context information sources and representation. The aim of this paper is to propose Jini [2] as technology which may be considered for *ContextWare* implementation. The detailed analysis of *ContextWare* requirements is compared with functionality of the services provided by Jini. These considerations are illustrated by the description of prototype *ContextWare* implementation. This way the mapping of *ContextWare* architecture services to customized Jini services is specified and validated. The missing element of this mapping concerns the aspect of context information aggregation and identification across different administrative domains. This requires semantic specification

P. Havinga et al. (Eds.): EUROSSC 2006, LNCS 4272, pp. 54–66, 2006.

of context data. The existing solutions are based on modeling semantic using ontology [4]. The paper is proposing extension of *ContextWare* architecture with context ontology database. This database may be referred during context information aggregation process and for resolution existing data representation conflicts. The proposed approach enhances the context dissemination scenario.

The structure of paper is as follows. In Section 2 *ContextWare* architecture and its requirements is presented. Section 3 presents concept of this architecture implementation with Jini that is related to DHT based solution. The mapping of *ContextWare* building blocks to Jini services is described. Section 4 discusses the *ContextWare* architecture extension with components providing semantic interoperability supporting context aggregation. Finally Section 5 contains description of prototype implementation of context dissemination and management system.

2   Requirements Context Dissemination and Aggregation System

AN networks context has been used [11] to enhance human-computer and computer-computer interaction, thereby helping achieve the vision of providing seamless computing and networking anywhere, anytime. Dey [9] provides the following key definition: *"context is any information that can be used to characterize the situation of an entity. An entity is an object, place or person that is considered relevant to the interaction between a user and an application, including the user and applications themselves."*

The overall *ContextWare* (CW) architecture, proposed for context dissemination and aggregation for AN is proposed by [10,23] and depicted in Figure 1. In essence CW is a mediator between context sources and sinks and its design resulted in a flexible architecture whose building blocks can be distributed to address scalability while breaking down complexity as the system usage and size grow. The scope of CW ranges from a single context lookup, through context associations (stable context exchange), to context dissemination strategies being defined dynamically, and finally to purposeful context processing, e.g. aggregation of network level context information. It represents an approach to network context handling that combines already known advances in the area of context-awareness support in restricted end-user single-administrative domains with new techniques for extending the exploitation of context-awareness to network control functions and across administrative domains.

The main building component of CW architecture is Context Coordinator (ConCoord) which is introduced to meet the requirements on creating a gateway into the *ContextWare* architecture and deals with indexing, registering, authorizing and resolving context names into location addresses. The second component, Context Manager (CM) is created to meet the requirements on creating appropriate Context Associations between clients and sources, managing the contents of the Context Information Base (CIB).

Context Sources and Context Managers are registered with ConCoord to be looked up by Context Consumers any time their contact information. Context information is modeled as data objects identified by Universal Context IDs (UCIs). Given the nature

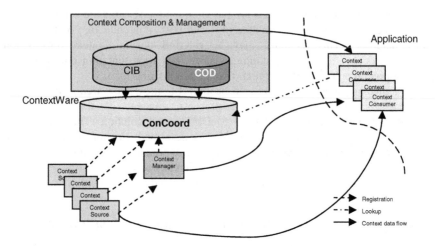

Fig. 1. *ContextWare* architecture

of context information, its sources and its potential clients, it is envisaged that creating a system that scales well is of primary importance.

ConCoord provides functionality necessary to map UCIs to contact information on demand. The contact information provided by ConCoord could be as simple as a string containing IP address and port number, or more complex as a proxy object implementing access protocol to Context Source, e.g. SNMP client.

This general scenario CW activity could be further enhanced with semantic information necessary for context aggregation and resolving inconsistency with data representation existing between different AN. As the process of AN composition is the key concept of this technology we propose to equip CW with Context Ontology Database (COD). The role of this component will be described in more details in Section 4.

The requirements regarding CW architecture could be summarized as follows:

1. Context Sources should be localizable in same way e.g. via discovery protocol, repository implemented as DHT or lookup service.
2. Context Sources could be implemented using different technology, and may communicate using different protocol with the context consumer.
3. Direct source to consumer response model of context data communication should be provided; it means that the context source forwards data directly to the context consumer.
4. Context Consumers should be dynamically equipped with a suitable proxy providing communication to the selected context source with suitable protocol and communication pattern.
5. Context Sources could be grouped and combined in one complex context data by a Context Manager. The Context Manager should look like a standard Context Source for a Context Consumer.

6. The source of context may appear or disappear in any moment of time, so interested Context Consumer should be able to be aware of this process. The subscription and triggering could be of some use for this purpose.
7. ConCoord implementation should be scalable, distributed, fault-tolerant, and secure.

These requirements on CW are guiding points for selection and proposition of appropriate implementation technology.

3 Implementation of *ContextWare* Architecture

CW architecture proposition does not rely on any particular implementation technology. The heterogeneity of Context Sources and Consumers implementation is rather obvious fact of AN nature, but ConCoord and CM implementation technology should be very carefully selected.

3.1 Implementation Technologies

Many papers propose usage of the distributed hash table (DHT). DHT [21,22,23], is a distributed system that provides a traditional hash table's simple put/get interface using a peer-to-peer overlay network. At the most basic level, it allows a group of distributed hosts to collectively manage a mapping from keys to data values, without any fixed hierarchy, and with very little human assistance. This building block can then be used to ease the implementation of a diverse variety of peer-to-peer applications such as file sharing services, DNS replacements, web caches, etc. Taking these features into account DHT is becoming interesting technology to use for some elements of *ContextWare* architecture implementation. The most natural concept of DHT application is implementing ConCoord with this technology. ConCoord provides the mapping from UCI (Uniform Context Identifier) to contact information.

Instead in our prototype implementation of CW we decided to use Jini. Jini was successfully used in research for service discovery in context-aware middleware   in mobile and wireless networks [25, 26]. Jini provides a functionality, which is required by *ContextWare*, as follows:

- Provides LookupService, which is a Jini service registry with lookup mechanism. LookupService may be distributed, and it may be found via LookupDiscoveryService. Services are searched by properties defined by user defined properties (e.g. UCI);
- All services in Jini are represented in LookupService as proxies. These proxies are written in Java language, however may communicate with services using service specific protocol, e.g. if service speaks with SNMP protocol, proxy must provide communication with service via SNMP;
- Dynamically equips a client with necessary proxy object which is provided as a contact information. The only thing a client must know, is an information to look up in the service; e.g. service interface and/or some user-defined properties for the service;

- Provides LeasingService. It assures that only up-to-date services exist in LookupService. Service must renew its lease in specified amount of time (e.g. one minute). If it is not done, service is removed from LookupService (e.g. in case of service or network crash);
- Offers support for remote events, which are required for *push* model of communication (Context Client subscribes for notifications about context changes). Remote events also use Leasing mechanism e.g. Jini LookupService may notify some clients about new services appearing in the network and about their disappearing.

The comparison of Jini provided functionality with previously discussed DHT is summarized in Table 1.

Table 1. Comparison Jini and DHT usage

Required functionality	Jini	DHT
Context source localization	Supported	Supported
Context source proxy dynamic provisioning	Supported	Not supported
Context source avaliability monitoring	Supported	Not Supported
Context source avaliability notification	Supported	Not supported
Support for composition of context space	Not yet defined	Proposed
Support for binding heterogenoius context sources	Supported	Not supported
Distribution of binding information	Limited	Full
Fault tolerance	Low	High
Security	Could be added	Difficult to add
Deployment process complexity	Low	High

Taking into account only this functional analysis, it is evident that Jini offers very attractive environment to be chosen for CW implementation. This conclusion should be further justified by performance study and scalability study but this aspect is out of scope of the presented paper.

3.2 Mapping *ContextWare* to **Jini**

ContextWare architecture building blocks may be mapped to Jini programming model in the following way :

- Context Manager is mapped to Jini Service. This service provides methods for context management:
  - get() (gets context managed by Context Manager);
  - subscribe() (subscribes for notifications about context changes);
  - sourceAdded() (adds Context Source to Context Manager);
  - sourceRemoved() (notifies CM, that specified CS is no longer available in AN; this event is executed, when ConCoord notices, that CS is no longer renewing its lease);

- Context Source is implemented as Jini Service which provides methods for context dissemination: `get()` (gets context from this source), `subscribe()` (subscribes for notifications about context changes);
- ConCoord is mapped to Jini LookupService; When CM or CS appears in Ambient Network, it registers in LookupService and become available for other entities in AN or clients. Primitives for Context Coordination are mapped to LookupService methods:
    - `register()` (registers Context Source or Context Manager in ConCoord) is mapped to joining Jini LookupService;
    - `resolve()` (resolves CSs or CMs location) is mapped to searching LookupService;
- Context Client is represented as Jini Client. It is prepared as a PAP module (described in Section 5). Context Client contains method `notify()` (which notifies client about context changes), which is mapped to remote Jini event.
- Context Description; It is mapped to OWL Ontology (described in Section 4).

The proposed mapping shows that Jini programming model in a very natural way satisfies most of context *ContextWare* architecture implementation requirements. Due to distributed architecture, notification and leasing mechanisms Jini assures elements of fault tolerance and auto-configuration. Moreover, Jini is a lightweight framework, designed especially for mobile devices, so it looks like ideal solution for Ambient Networks.

4   Interoperability Aspects

The most of the context related research [3, 4, 5] select tree a good data structure to represent context information. It allows for generalization of context information processing issues and provides support for aggregation of different contexts data, and is suitable for reasoning mechanisms. The tree representation corresponds also directly to ontological descriptions of context information. This important aspect will be further elaborated in this section.

4.1   Ontologies

Ambient Networks [1, 23] have identified the use of a common ontology as a requirement for entities (Functional Entities) to exchange context information in a consistent and unambiguous way. Compositions/decompositions taking place in Ambient Networks require efficient knowledge transfer and automatic decision making. This in turn demands proper conceptualization of the context domains having roles in the process of composition/decomposition.

That is why an ontology language [6] is a good language for defining context. Ontologies are often used to support semantic web [12], and may be adopted for context description. Ontolgies are often used in context management systems, in ubiquitous computing [27,28], because of their flexibility.

There are few ontology language specifications, like DAML+OIL [13], KIF [18], OCML [19], F-logic [20], which are specific for different issues.

We decided to use Web Ontology Language (OWL [8]), standard recommended by W3C. OWL is represented by XML/RDF document, which may be read by any XML/RDF-enabled tools [14,15]. There are also reasoners for OWL, like [16,17].

According to Ambient Networks issues, following requirements for *ContextWare* are met by OWL:

- Conceptualization requirements:
  - formality;
  - efficient reasoning and inference mechanisms;
  - dynamic creation, modification, and extensions of the ontology model;
  - models merging, checking and partial knowledge validation;
  - support for interoperability between different models;
- Engineering requirements
  - ease of development of new ontologies for different domains in ANs;
  - upgrading existing ontologies / ontology evolution;
  - integrating / merging existing ontologies;
  - compatibility with standards;
  - representation language independence.

This rather general consideration can be illustrated by a simple example. This is only initial, sample ontology, but it may be extended in any way. This article shows, how ontologies are used, not the complete ontology. Let us consider situation, where there are few Context Sources and a Context Manager, which is operating as a Context aggregator. Using ontologies we are able to aggregate context from these sources. New ontology is the union of the source ontologies. The merged ontology captures all the knowledge from the original ontologies [24]. It could be done in a semi–automatic

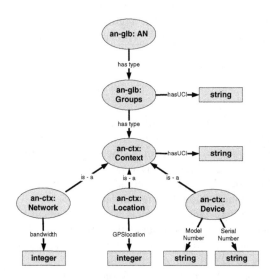

Fig. 2. Simple ontology

way. Of course there is a problem in case of ontology conflicts (if few context sources provide context which overlap), however there is an idea to create Ambient Networks Ontology, to avoid this issue. Fig. 2 depicts simple ontology prepared for testing model (demo is described in Section 5).

The whole ontology provided by AN (with all Context Sources and Context Managers), is stored in Context Ontology Database. This ontology should be defined for every Ambient Network and stored in an ontology repository, as depicted in Fig. 1.

4.2 Universal Context Identifier

Context Sources and Context Managers are registered in ConCoord using UCI, which is a new type of Universal Resource Identifier proposed in [1, 23]. Fully qualified UCI looks as follows:

```
ctx://domain.org/path?options
```

where:

`ctx`	– URI scheme for context
`domain.org`	– DNS domain name within which the context object exists
`path`	– the sequence of words separated by slashes
`options`	– specification of optional modifiers like data encoding.

If client or source wants to use current domain, it may use simplified UCI format:

```
ctx:/path?options
```

Path part in the UCI defines detail level for specified context. If client wants to get more general context, it asks for shorter path. If client needs more detailed context, asks for UCI with longer path. For example if there are two devices in AN with following UCIs:

```
ctx://www.ambient-networks.org/ambientnetworks/devices/laptop
```

```
ctx://www.ambient-networks.org/ambientnetworks/devices/pc
```

and client asks for UCI:

```
ctx://www.ambient-networks.org/ambientnetworks
```

he gets context for whole AN. However, if he asks for UCI:

```
ctx://www.ambient-networks.org/ambientnetworks/devices/laptop
```

he just gets context for laptop device.

Client may ask ConCoord only for UCIs that are registered and bound with Context Source or Context Manager. For instance with:

```
ctx://www.ambient-networks.org/ambientnetworks
```

must be associated Context Manager to perform context aggregation. This CM gets context from connected sources and sends it to the client. In our example CMs are used for aggregation of context from Context Sources and other CMs. Connected CMs forms structure of a tree, as depicted in Fig. 3.

Context Managers may also provide additional functionality, like caching context fetched from source to reduce network load. Important thing is, that depicted tree is rebuild automatically when new CM appears in the network or disappears (it may happen because of a very dynamic Ambient Networks structure).

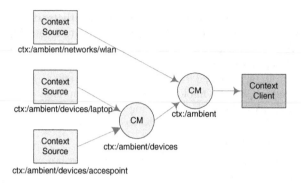

Fig. 3. Tree of context associations in AN with two CMs

Universal Context Identifier is used for locating the Context Source or the Context Manager within AN. It is used as a parameter of the ConCoord interface operations described before. Here more details on these operations are showed:

- register(UCI, contact)

  registers in the ConCoord UCI with contact to specified Context Source or CM (in Jini we use service proxies);

- resolve(UCI)

  returns the contact (proxy) for CS or CM registered in ConCoord, which provides context with specified UCI.

4.3   Ontology and UCI Relationship

UCI is strongly connected with ontology provided by Context Source/CM with specified UCI, as shown in the Fig. 4. Depicted context tree is an instance of ontology

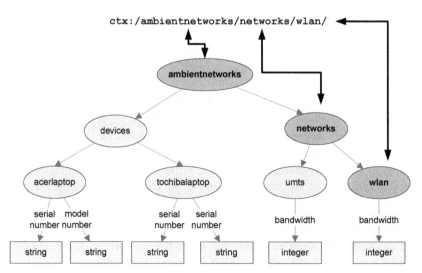

Fig. 4. Simple context tree for Ambient Networks

class mentioned before. Black arrows show relations between UCI and chosen ontology instances.

Each part of the UCI has a corresponding node (instance of ontology class) in the context tree. This way ontology could be considered as template for UCI.

5 Case Study – Demo

To verify our point of view on context dissemination and aggregation, we decided to build prototype of *ContextWare* architecture which would prove our concept. We designed a simple ontology as our prototype. It was necessary because of lack of a common AN ontology. Ambient Networks has identified the use of a common ontology as a requirement to exchange context information in a consistent and unambiguous way.

Ambient Networks describes approach for the development of the integrated prototype in Phase II of the project, which is mainly concerned on implementation. For this purpose *Joint Use Case* scenario was described. Scenario consists of seven scenes. Main plot of this scenario is one day of a group of peoples which goes on the trip to Liverpool. In the meanwhile, these peoples extensively use electronic devices which are all connected by Ambient Networks. We chose for our purpose only three scenes, the ones which deal with context awareness a lot. Action takes place in a railway station and later in the train.

Jini was chosen for development of ContextWare mechanism due to advantages described earlier. In phase I of the AN project PAP [1] (P2P ACS Prototype) has been

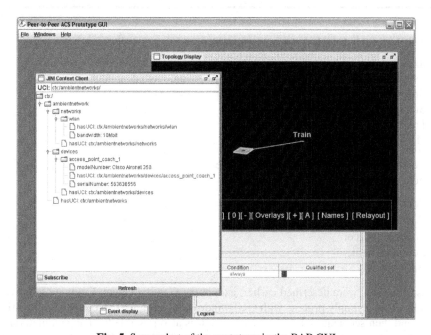

Fig. 5. Screenshot of the prototype in the PAP GUI

developed in order to demonstrate P2P network management concepts, including policy based network composition and hierarchical management overlays. It is a good integration platform for testing and developing new concepts. It allows to familiarize with the problems which may occur during final implementation. Fig. 5 depicts PAP GUI with few windows. *Topology Display* window illustrates details how Ambient Networks are composed together. All messages and logs are presented on *Event Display* window. To verify policies responsible for AN composing we can use *Policy viewer* window. PAP could be extended by adding new modules, as it was done in our case. *Jini Context Client* role is described below in details.

For context dissemination we need Context Sources, and for aggregation purposes we need Context Managers. In our scenario every electronic equipment has embedded a Context Source, and every place which is in the scenario, has Context Manager to provide information on it. All of the existing Context Sources in our prototype are presented in Table 2.

Table 2. List of existing ContextSources

	Places		**Actors**	
	Train	Railway Station	Bob	John
Context Sources	Network: WLan Network: WiMax	Networks: UMTS	Device: laptop	Device: laptop

Scenario which is used to verify our Jini based prototype has the following steps:

1. John is going to the Railway Station.
2. John is waiting for a train on the Railway Station.
3. Bob joins John.
4. Both men embarking the train.
5. Train leaves the Railway Station.
6. Men go to the next coach.

To write simple prototype of ContextWare architecture we had to develop all of the modules described earlier, that is:

- ConCoord;
- Context Manager;
- Context Source;
- Context Client;
- Ontology.

These elements were described earlier, except the Context Client. Context Client was developed as a simply process with GUI window in which we can query of a specified UCI, and then see acquired context as a tree. It is also possible to change a method of acquiring context by using push/pull method, by checking *Subscribe* box. Prototype has been tested on four laptops. PAP was extended by Jini LookupService. On each laptop an enhanced version of PAP has been installed. Each time, when some actors appear/disappear in the scene, the registration of Context Sources UCI with LookupService has been changed accordingly. It was visualized by PAP. Also Context Sources were started when actors location has changed. Fig. 5 depicts the Jini Context Client window. Context presented in the window is the answer of *ctx:/ambientnetworks/* query. As we can notice, in the Ambient Network in the Train, there is available WLan and its AccessPoint.

The implemented prototype shows that Jini programming model perfectly satisfies the *ContextWare* architecture implementation requirements. The prototype has been implemented with a very little implementation effort due to built-in Jini mechanisms that perfectly fit our needs.

6 Summary

The implementation of ContextWare architecture requires highly dynamic distributed environment. The heterogeneity of technology which could be used for context data generation makes the process of its dissemination and aggregation even more complex. Using DHT for context information distribution does not resolve many important problems such as dynamic appearing and disappearing of context sources or diversity of the context generation technology. The presented study shows that Jini fulfills most implementation and functional requirements of ContextWare. Jini programming model assumes heterogeneity of services implementation and allows the usage of any specific client-server communication protocol. This is particularly important for Context Managers implementation.

Jini Lookup-service provides functionality which could be easily exploited to specify context data with UCI and to represent context sources life cycle. It also fits well to requirements set by ontological representation of context data. With comparison to purely DHT based approach Jini offers much more open and flexible solution. The only problem is that Jini usage leads to rather more centralized ConCoord implementation. It could be eliminated by Jini LookupService replication and organization in federation. This particular aspect seems to be very important for AN and requires more detailed study. The possible solution may lead to DHT based implementation of ConCoord and Context Managers implemented with Jini.

References

[1] WWI-AN Ambient Networks Project / Reports, www.ambient-networks.org
[2] http://www.jini.org/, Jini Framework Home Page
[3] Coutaz, J., Crowley, J., Dobson, S., Garlan, D.: *Context is key*, Communications of the ACM 48(3). March 2005
[4] Kamienski, C., Fidalgo, J., Sadok, D., Lima, J., Pereira, L.: *PBMAN: A Policy-based Management Framework for Ambient Networks*.
[5] Gu, T., Wang, X.H., Pung, H.K., Zhang, D.Q.: *An Ontology-based Context Model in Intelligent Environments*
[6] Strang, T., Linnho-Popien, C., Frank, K., CoOL: *A Context Ontology Language to enable Contextual Interoperability*

[7] Uschold, M., Gruninger, M.: *Ontologies: Principles, methods, and applications.* Knowledge Engineering Review 11 (1996) 93-155

[8] http://www.w3.org/TR/owl-ref/ OWL Web Ontology Language Reference

[9] Dey, A.K., Abowd, G.D., Salber, D.. *A conceptual framework and a toolkit for supporting the rapid prototyping of context-aware applications.* Human-Computer Interaction, 16, 97-166, 2001.

[10] Balakrishnan, D., El Barachi, M., Karmouch, A., Glitho, R.: *Challenges in Modeling and Disseminating Context Information in Ambient Networks.* 2nd International Workshop on Mobility Aware Technologies and Applications (MATA 2005), Montreal, Canada, October 2005.

[11] R. Giaffreda, A. Karmouch, A. Jonsson, A. Karlsson, M. Smirnov, R. Glitho, and A. Galis. *Context-Aware Communication in Ambient Networks.* Wireless World Research Forum 11th Meeting, Oslo, Norway, June 2004.

[12] http://www.w3.org/2001/sw/ W3C Semantic Web Home Page

[13] McGuinness, D. L., Fikes, R., Hendler, J., Stein, L.A.: *DAML+OIL: An Ontology Language for the Semantic Web.* In IEEE Intelligent Systems, Vol. 17, No. 5, pages 72-80, September/October 2002.

[14] http://jena.sourceforge.net/, Jena – A Semantic Web Framework for Java Home Page

[15] http://protege.stanford.edu/, Protégé ontology editor and knowledge-base framework Home Page.

[16] http://www.racer-systems.com/, RacerPro ontology reasoner Home Page

[17] http://fowl.sourceforge.net/, F-OWL: An OWL Inference Engine in Flora-2 Home Page

[18] Ginsberg, M.L.: *Knowledge interchange format : the kif of the death.* Technical report, Stanford University, 1992

[19] Motta, E.. *An Overview of the OCML Modelling Language.* In Proceedings of the 8th Workshop on Knowledge Engineering Methods and Languages (KEML'98), 1998

[20] Kifer, M., Lausen, G.: *F-logic: A higher-order language for reasoning about objects,* inheritance, and scheme. In Proceedings of ACM SIGMOD Conf. on Management of Data, volume 18, pages 134 - 146, Portland, Oregon, June 1989

[21] Dabek, F., Li, J., Sit, E., Robertson, J., Kaashoek, M.F., Morris, R. *Designing a DHT for low latency and high throughput.* Proc. 1st Symposium on Networked Systems Design and Implementation (NSDI 2004)

[22] http://bamboo-dht.org/, The Bamboo Distributed Hash Table

[23] Galis, A., Ocampo, R., Jean, K., Cheng, L., Karmouch, A., Samaan, N., Harroud, H., Giaffreda, R., Dang, J., Kanter, T., Reichert, C., Glitho, R., El Barachi, M., Belqasmi, F., Bałos, K., Szydlo, T., Szymacha, R., Zielinski, Z.: *ContextWare Infrastructure for Ambient Networks*, submitted on 4th June 2006 to IEEE Wireless Communications Magazine

[24] de Bruijn, J., Ehrig, M., Feier, C.: *Ontology mediation, merging and aligning,* May 2006

[25] Bellavista, P., Corradi, A., Montanari, R., Stefanelli, C.: *Context-aware Middleware for Resource Management in the Wireless Internet.* IEEE Transactions on Software Engineering ,vol. 29, no. 12, pp. 1086-1099, December, 2003.

[26] Cummins, S., O'Reilly, F.: *Service Delivery in Wireless Ad Hoc Networks using Jini.* M-Zones White Paper May 03, white paper 05/03, Ireland, May, 2003

[27] Christopoulou, E., Goumopoulos, C., Kameas, A.: *An ontology-based context management and reasoning process for UbiComp applications.* In Proceedings of the 2005 joint conference on Smart objects and ambient intelligence, pages: 265 - 270, Grenoble, France, October 2005,

[28] Ranganathan, A., Campbell, R.: *An infrastructure for context-awareness based on first order logic.* Personal and Ubiquitous Computing. 7(6):353-364, 2003.

Discovery and Composition of Services for Context-Aware Systems

Cristian Hesselman[1], Andrew Tokmakoff[1], Pravin Pawar[2], and Sorin Iacob[1]

[1] Telematica Instituut, The Netherlands
[2] University of Twente, The Netherlands
{cristian.hesselman, andrew.tokmakoff, sorin.iacob}@telin.nl,
p.pawar@utwente.nl

Abstract. We consider the challenge of dynamically adapting services to context changes that occur in ubiquitous computing environments (e.g., changes in a user's activity) and propose the Context-Aware Service Enabling (CASE) platform for that purpose. The CASE platform combines context-aware service discovery with service composition, acting as an enabler for the development of adaptive context-aware applications. In this paper, we illustrate the need for context-aware service discovery and composition in pervasive 4G environments and present the architecture of the CASE platform. The CASE platform enables applications to easily adapt to changes in service availability, which may result from changes in client and/or service context. We also provide an overview of the platform's technical realization.

Keywords: Service Discovery, Context-awareness, Service Composition.

1 Introduction

The vision of 4G mobile networks is one of new high-speed radio access network technologies, an all-IP network, and a ubiquitous service platform [3]. Such a service platform has a pivotal role in the 4G vision [8] as it provides a common underlying set of functions that are enablers for the realisation of innovative new services. These services are expected to make use of advanced mobile terminals that have various radio technologies at their disposal and will be made available by providers that build upon the core services offered by the underlying service platform.

Two essential aspects of ubiquitous computing, as seen in the 4G vision, are those of service discovery and of context-awareness. The first is a service that allows applications and/or services to discover and bind to services that appear and disappear in highly dynamic mobile environments. The discovery of appropriate services can benefit from knowledge of both client and service context. Furthermore, it may be the case that no direct match for the requested client service can be obtained. In this case, service composition can be utilised to dynamically "construct" a composite service that matches the request of the client. This may also be subject to changing context since a composed service may need to be re-composed over time or may become "inappropriate" for a client as its context or that of the composed service changes.

P. Havinga et al. (Eds.): EUROSSC 2006, LNCS 4272, pp. 67–81, 2006.

In the following sections, we will further discuss some of these essential concepts and present them in relation to the Context-Aware Service Enabling (CASE) platform, which provides a suite of core functionalities for context-aware service discovery and composition that are needed as part of a broader '4G services platform'. We first introduce our main concepts (Section 2) and then present the architecture of the CASE platform (Section 3). Next, we provide an overview of the platform's technical realization (Section 4), which is currently under development. We conclude with a discussion of related work (Section 5) and a summary (Section 6).

2 Context-Driven Service Adaptation

One of the most critical issues in pervasive computing environments is that of ever-changing context. This applies equally well to mobile devices as to the (fixed) services these devices use: mobile devices are regularly subject to location, network, and power context changes, whilst services can for instance be subject to changes in the types of devices they have to serve. In pervasive computing environments, such changes should result in a service being dynamically adapted to the new context of a mobile device or of the service itself.

2.1 Motivating Example

In Fig. 1, we present an example of an adaptive service that streams the latest news (audio and video) to mobile users over the Internet. The service can dynamically add and/or remove the video stream from a multimedia news transmission in reaction to changes in the context of a user.

In Fig. 1, the service responds to changes in the activity of user Bob. Bob's initial activity is 'driving on highway 35' (Fig. 1a), which results in the service delivering the news transmission in audio-only mode (for safety reasons). Bob then stops at his destination and alights from his car. This has the effect of changing his activity to 'walking in downtown Amsterdam' (Fig. 1b). As a result, the service adjusts to also deliver the video part of the transmission to Bob.

This change in service configuration could also result from other context changes, such as a change in available networks. Similarly, it is also possible that the service's context may change (e.g., due to network outages or overloading) which would also result in an adapted service being delivered to Bob. In the remainder of this section, we provide a more detailed discussion of some important underlying concepts, including services, context sources, context agents, and service adaptation.

2.2 Services

We consider a *service* to be a unit of well-defined functional behaviour (in syntax and semantics) that is offered by a software entity for use by other software entities. The adaptive broadcasting service of Fig. 1 is an example of such a service.

A service can be a *composite service* in that it can consist of one or more *constituent services*. A constituent service provides part of the composite service or helps other constituent services to do so. In general, a constituent service can itself be

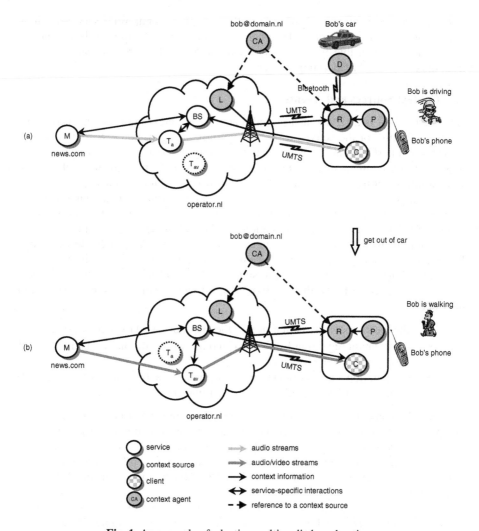

Fig. 1. An example of adaptive multimedia broadcasting

a composite service, which means that it can also be further decomposed into yet another set of constituent services. In the example of Fig. 1a, the broadcasting service (BS) is a composite service. Its constituent services are a multimedia streaming service (M) operated by a newscaster (news.com) and an audio transcoding service (T_a) operated by an UMTS operator (operator.nl). The streaming service transmits an audio stream, which the transcoding service receives and adjusts to match the capabilities of Bob's UMTS phone (e.g., by scaling the stream to a lower bit-rate).

Constituent services are arranged into a *service graph*, where the graph's edges indicate how the services relate to each other. Different constituent services of the same composite service could potentially operate in different execution domains. In Fig. 1a, the constituent services run in both the news.com domain (M) and also in the domain of the UMTS operator (T_a).

Clients are applications that interact with (potentially composite) services. These interactions are service-specific. The client shown in Fig. 1 (C) runs on Bob's mobile phone and is responsible for rendering the audio and video streams that it receives from the composite broadcasting service.

2.3 Context Sources and Agents

A *context source* is a service that provides access to context information, such as the location of a user or the activity a user is currently engaged in [1]. A context source provides an interface that enables clients to directly access context information via a request-response interaction or by subscribing to events that signal a change in context information (e.g., when a user moves from one room to another).

As with all services, a context source can be a composite context source that can aggregate context information and also determine higher-level context information from more elementary context information (cf. the interpreters of [16]). Fig. 1 depicts an example in which a composite context source (R) enables a client to determine Bob's current activity (e.g., 'Bob is driving on highway 35' or 'Bob is walking in downtown Amsterdam'). To be able to supply this sort of information, the composite context source consists of three different constituent context sources. They can provide:

- the location of a particular user in the operator's UMTS network (L),
- information regarding Bob's car (D) (e.g., who is driving it, its direction of travel and its velocity), and
- acceleration and orientation information about Bob's phone (P) (e.g., using a gyroscope and an accelerometer).

Using the context information provided by these constituent context sources, the composite context source can determine what activity Bob is currently engaged in.

Fig. 1 illustrates that composite and constituent context sources may operate in different execution domains. Context sources P and R reside on Bob's mobile phone, whereas L and D are located in the UMTS infrastructure and in Bob's car, respectively.

A *context agent* is a service that stores references to context sources. A context agent represents an *entity* whose context needs to become discoverable. These entities can be classified into people, places (e.g., a meeting room), and things (e.g., mobile devices or software components) [16]. The references that a context agent stores point to context sources that can currently provide context information about the entity represented by the context agent. For example, the context agent associated with the person 'Bob' could contain a reference to the composite context source that provides access to Bob's current activity (R) and a reference to the context source that provides lower-level information about Bob's current location (context source L).

A context agent acts as a single, persistent point of access for context information about a particular entity and should therefore be 'always on'. Each context agent has a unique identifier, for instance based on the type of URLs defined by the Session Initiation Protocol (SIP) [18]. In this case, the context agent of Bob (Fig. 1) would be identified by a SIP URL like sip:bob@domain.

After a client has resolved the identifier of a context agent to a network address (e.g., using SIP and DNS), it can access the context agent. A context agent provides a request-response interface, which enables clients to retrieve a subset of the references stored by the context agent. A request indicates the types of context sources the client is interested in (e.g., context sources that can provide information about the temperature at Bob's current location). The context agent's response consists of references to context sources that can provide this information. A context agent also provides a publish-subscribe interface, which enables clients to asynchronously receive updates in the context agent's list of context sources. After a client has obtained a set of references to context sources, it can use their interfaces to get the actual context information.

A context agent can be realized in various ways, for instance as a web server [17]. A context agent can furthermore be combined with a context source, in which case the context agent also implements a context source interface. This means that the context agent can also return actual context information instead of just references to context sources. In this case, a context agent is similar to the context aggregators discussed in [16].

Observe that a context agent does not need to be physically co-located with the entity it represents. For example, the context agent of a mobile device could be located somewhere in the network infrastructure rather than on the device itself.

2.4 Service Adaptation

In pervasive computing environments, a service may need to be adapted in response to a context change. These adaptations may need to occur *while the service is being used*. In the example of Fig. 1, the broadcasting service (BS) is adapted in response to a change in Bob's activity (from 'driving' to 'walking') while Bob is listening to/watching a news transmission. This dynamic adaptation is realized by a re-composition of the service's graph: the audio transcoding service (T_a) is removed and is replaced with a transcoder that can handle both audio and video streams (T_{av}).

Context sources provide the means to detect context changes, but may themselves need to be re-composed as a result of such a change. For example, when Bob gets out of his car (Fig. 1), the composite context source (R) changes since the car's context source (D) becomes unavailable. In the example, context source R can still function without D, but D may also need to be replaced with an equivalent context source.

3 Service Discovery and Composition

The main function of the CASE platform is to dynamically adapt services (including context sources) by changing their composition in response to context changes, possibly while these services are being used (cf. the example of Fig. 1). To accomplish this, the platform consists of a composition service and two types of discovery services: a context-aware discovery service and a basic discovery service. Fig. 2 illustrates this. The arrows in Fig. 2 represent interactions.

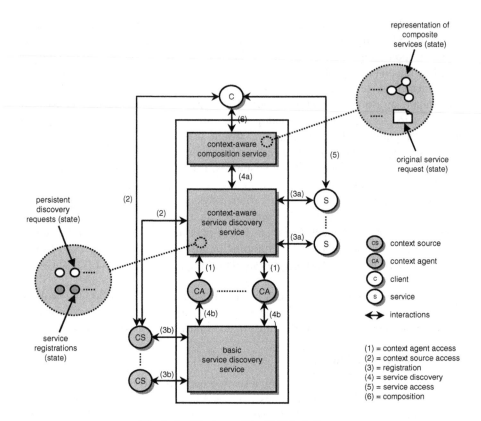

Fig. 2. Architecture of the CASE platform

The CASE composition service dynamically (re)composes services based on requests from clients and returns references to composite services to these clients (interaction 6 in Fig. 2). Clients can subsequently access these services (interaction 5).

The composition service uses the context-aware discovery service to locate the constituent services its needs for a particular composition (interaction 4a). Clients may however also bypass the composition service and directly interact with the context-aware discovery service.

The context-aware discovery service dynamically discovers services. It optimizes the discovery process by means of context information (e.g., by only considering near-by services [2]), which it obtains via context agents (interaction 1 in Fig. 2). Context agents provide references to context sources (see Section 2.3), which they find by accessing the basic service discovery service (interaction 4b in Fig. 2). This discovery service is context-unaware and can be implemented using well-known discovery protocols such as SLP or WS-Discovery. The context-aware discovery service access the context sources to actually get the context information it needs (interaction 2).

Observe that the distinction between a context-aware discovery service and a basic discovery service is a logical one. In an implementation, the two discovery services may partly overlap. Also note that in a pervasive computing environment the composition service and the two discovery services will typically be realised in a distributed manner.

In this paper, we will concentrate on the context-aware service discovery service (Section 3.1) and in particular, on the interactions that occur at its interfaces. We will also briefly discuss the composition service (Section 3.2) and its interfaces and interactions.

3.1 Context-Aware Discovery Service

The context-aware discovery service is an extension of a traditional discovery service in that it uses context information during discovery. The discovery service obtains this information through context agents, which we introduced in Section 2.3. Fig. 3 shows an example in which the discovery service makes use of three context agents, one associated with Bob, one with the service S, and one associated with Bob's car. Each context agent stores references to context sources that can provide context information about the associated entity (Bob, S, and Bob's car in this example). The context-aware discovery service obtains context information in two steps: it first obtains references to relevant context sources through one or more context agents (interaction 1 in Fig. 2/Fig. 3) and then accesses those context sources to obtain the actual context information (interaction 2 in Fig. 2/Fig. 3).

In this paper, we assume that context agents can be found through their identity (e.g., a SIP URL) using an external discovery mechanism (e.g., SIP and DNS). We also assume that context agents are able to deal with changes in their set of context sources and are able to keep this set current.

As with established discovery services [7], the CASE context-aware discovery service provides three interfaces:

- A *registration* interface, which enables services to become discoverable by *registering* their descriptions with the discovery service;
- A *discovery* interface, which allows discovery clients (the composition service or the clients of the platform) to *find* services by matching their discovery requests with the descriptions of registered services. During discovery, a client obtains information about the existence of services, their applicable parameters, and their semantics (e.g., using ontologies [4]).; and
- A *bootstrapping* interface, which clients and services use to discover the service discovery service.

In Fig. 2 and Fig. 3, the interactions that occur at the registration and discovery interfaces are labelled 3 and 4, respectively. We will not consider the bootstrapping interface any further in this paper.

The interfaces of the CASE discovery service extend traditional registration, discovery, and bootstrapping interfaces. In this paper, we define these extended interfaces in terms of a set of device-local service primitives, their parameters, and the order in which the primitives are exchanged. To keep the discussion as general as possible, we assume that service primitives are exchanged asynchronously.

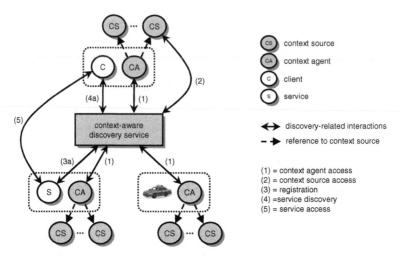

Fig. 3. Service discovery example

The discovery interface supports active and passive discovery. In *active discovery*, discovery clients actively request the discovery of certain services, whereas in *passive discovery* they wait for the discovery service to push such services to them (on a subscription basis). Passive discovery is particularly useful when a discovery client is constantly looking for 'better' services. Passive discovery might for instance be useful in the scenario of Fig. 1 if Bob's client is continuously looking for transcoding services that can deliver a certain new transmission at the highest possible quality in Bob's current context.

Active Discovery Interface. The active discovery part of the discovery interface consists of a discovery request primitive and a discovery response primitive (interaction 4a in Fig. 2/Fig. 3). As in traditional service discovery, clients use a discovery request to invoke discovery and subsequently receive a response that contains references to matching services. The request contains the usual parameters, which are a semantic specification of the services the client is trying to find (e.g., transcoding services), a set of constraints (e.g., transcoders that support MP3 audio), and a description of the scope in which the discovery service should look for matches (e.g., in terms of a geographical area or a number of network hops) [7].

The CASE-specific parameters in a discovery request are:

- A client context specification, which describes the context information of the discovery client. This parameter either consist of actual context information (which the client obtained via its context agent) or of a reference to the client's context agent. The client context specification is optional because some clients may not aware of having a context agent, and
- An (optional) set of additional constraints that describe the context that prospective services should to be in (e.g., printing services that must be located in a certain building).

Each of the references in a discovery response primitive comes with a description of the corresponding service, which enables the client to intelligently select the most appropriate service out of a number of alternative matches.

Passive Discovery Interface. The passive discovery part of the interface consists of three primitives: a persistent discovery request, a persistent discovery response, and a persistent discovery notification (also interaction 4a). A persistent discovery request is essentially an active discovery request that has a specified lifetime. Discovery clients use a persistent discovery request to instruct the CASE discovery service to generate a discovery notification when it discovers services that are 'better' than the ones it proposed in previous notifications. Before issuing such notifications, the discovery system first confirms the receipt of the discovery request by passing a persistent discovery response back to the discovery client.

With passive discovery, the scalability of the CASE discovery service is an important concern because it needs to maintain state for each outstanding persistent discovery request (see the enlargement in Fig. 2). The service discovery service therefore uses softstate persistent requests, which means that it removes the state associated with a persistent request unless that state is refreshed before a specified time (leasing).

The parameters of a persistent discovery request are similar to those of an active discovery request. The differences are that a persistent discovery request also contains:

- A reference to the client (e.g., in the form of a URL) so that the service discovery service can asynchronously deliver discovery callback notifications; and
- A specification of the types of discovery notifications the client wishes to receive (e.g., notifications that signal the appearance of a new matching services or events that indicate the disappearance of such as service).

A persistent discovery response indicates if the discovery service successfully handled the preceding request.

Registration Interface. The registration interface of the CASE context-aware discovery service is almost the same as for established service discovery services. The most important primitives are registration requests and registration responses (interaction 3a in Fig. 2/Fig. 3). A service uses a registration request to register with the discovery service, which then passes back a registration response.

The usual parameters of a registration request are a service description (augmented with semantic descriptions), a specification of the scope in which the service is available (e.g., a number of network hops or an administratively defined scope), and a self-reference so that discovery clients can actually contact the service.

The CASE-specific parameter of a registration request is the context of the service (optional), either in the form of the actual context information or as a reference to the service's context agent.

Operation. Fig. 4 describes the sequence of high-level operations used for processing persistent discovery requests. The diagram illustrates how context information

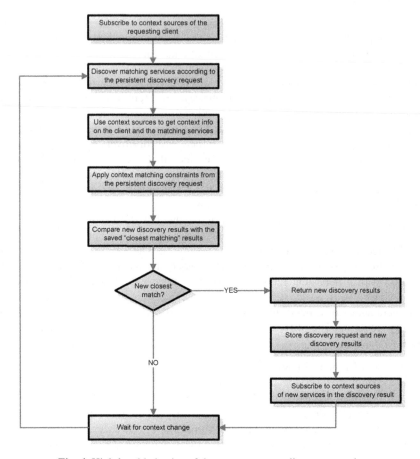

Fig. 4. High-level behavior of the context-aware discovery service

triggers the recalculation of a set of matching services. This may result in new services being returned to the client in discovery notifications.

3.2 Composition Service

The composition service is responsible for dynamically constructing composite services, based on client requests (see interaction 6 in Fig. 2/Fig. 3). The composition service can also re-compose the service to keep it "matched" with what the client initially requested if, for example, some of the service's constituent services suddenly become unavailable or when there is a context change (interaction 2, callback from a context source). In this case, the composition service proposes one or more recompositions of the service that the client is utilizing. The client can then decide which (if any) of the proposed composite services it would like to bind to.

In Fig. 1, the composition service could for instance propose a recomposition of the broadcasting service (T_a replaced by T_{av}) shortly after Bob gets out of his car. The client on Bob's mobile host can then decide if it wants to bind to the newly composed broadcasting service. Alternatively, the client could also delegate such binding decisions to the composition service, but this would require the client to inform the composition service of its 'binding policy'.

A service request to the CASE composition service includes a semantic construct parameter that specifies the desired functionality (i.e., the composite service). The composition service returns a reference to the composite service that it has constructed.

To automatically compose services, discoverable constituent services must not only provide an explicit description of their interfaces and parameters, but also of their functionality (see the registration interface in Section 3.1). A commonly used approach for functionality description relies on the use of domain Ontologies [4]. Such semantic service descriptions enable the design of effective mechanisms for automatic composition [5].

Assuming that the functionality of a (composite) service and the query for that service (see interaction 6 in Fig. 2/Fig. 3) are expressed in a semantically consistent way (e.g., in terms of the same ontology), it is possible to estimate the "gap" between the functionality of the requested (composite) service and that offered by any of the available (constituent) services [6]. Abstractly speaking, the behaviour of the CASE composition service can then be defined as an iterative process by which new functions (i.e., constituent services) are added at runtime to an existing aggregation (i.e., composite service), until a certain acceptable error threshold between the desired and available functionality is reached.

After each iteration, the composition service evaluates the newly constructed composite service and calculates a *utility measure* (u) as a function of the semantic similarity between the required and achieved functionality, and some non-functional constraints (e.g., response time and cost). The context of the client can influence the set of constituent services that the composite service selects for a particular composition as well as the way in which they are arranged in the composite service's service graph. Once a service has been constructed, context changes such as those of Fig. 1 might require the composition service to select new constituent services (e.g., transcoders) and use them to recompose the composite service (e.g., BS in Fig. 1).

4 Technical Realization

Fig. 5 shows the technical realization of the CASE platform, which we are currently developing. We are concentrating our efforts on the implementation of the context-aware service discovery service and will add the composition service at a later stage. Our basic discovery service is the Jini lookup service [14], which we selected as a technology for initial investigations. At this stage, we also have omitted context agents from our implementation and let context sources directly register with the context-aware discovery service.

The Jini infrastructure enables Jini services to register with Reggie [10] (the Jini lookup service) using its discovery and join protocols. In our implementation, the context-aware discovery service, other services in the Jini network, and context sources are all Jini services. A reference to a context source consists of a serviceID, which is generated by the Jini lookup service when a context source registers with it.

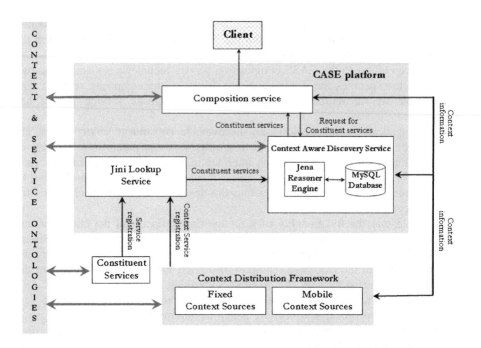

Fig. 5. Technical realization of the CASE platform

Services are registered using Jini's Service Entry functionality and provide references to their associated context sources. These context sources supply context information on the service according to the context ontology outlined in [19].

A context source uses the remote eventing mechanism provided by Jini to notify clients of changes in context information. A client (in our case the context-aware discovery service) interested in the context information implements a remote event listener interface to receive remote events. The context-aware discovery service also subscribes to the Jini lookup service so that it is notified when a new service registers.

As shown in Fig. 5, our implementation involves fixed context sources as well as mobile context sources. A mobile context source participates as a service in the fixed network using the Mobile Service Platform (MSP). The MSP design is based on the Jini Surrogate Architecture Specification [15], which enables devices that cannot directly participate in a Jini Network to join a Jini Network (with the aid of a third party). The MSP consists of an HTTPInterconnect protocol to meet the specifications of the Jini Surrogate Architecture and provides a custom set of APIs for building and running services on a mobile device.

A context source in the fixed network exports a service proxy to the Jini lookup service. The CASE context-aware discovery service uses this proxy to communicate with a context source. We use ontologies to describe context sources and services, thus facilitating a common semantics for context information and service descriptions (OWL-S). The Context Distribution Framework (CDF) provides the necessary APIs to implement context sources in the mobile and fixed network and to access context information. It also provides support to update and distribute context information using ontological representation.

Our implementation utilises Jena [9] for service and context matchmaking. Jena is a framework for building semantic web applications. It includes a rule-based inference engine, support for ontologies, a querying mechanism, and persistent storage capability using a database. We supply Jena with the context information obtained from context sources and utilize its querying capabilities to interpret context information for the purpose of matching registered services. When a client issues a persistent service discovery request, Jena stores the service context information in a MySQL database.

5   Related Work

This section provides an overview of existing approaches that use context information to assist in the service discovery and composition processes.

The work reported in [11] combines service-oriented and context-aware computing in order to provide composite services to users. Their service descriptions consist of service context information like location, usage conditions, and a Context Of Interest Function (COIF). During service discovery, the value of the COIF is calculated at run-time to select a better service if the matching process returns more than one service. The service composition process uses the client's context information to search for the closest basic services which support the user device. The set of services selected during service discovery is further refined using context parameters that are relevant for the composition. Our approach is similar to that described in [11]. However, the major factor which distinguishes the CASE platform is its support for dynamic service re-composition whilst a (stateless) service is being utilized.

The architecture discussed in [13] builds context-aware applications as a dynamically-composed sequence of calls to fine granularity Web services based on context information. The user specifies a request for a composite service that consists of context data and a goal. This goal is converted into sub-goals using a BPEL4WS control flow template. A goal-oriented planning system SHOP2 transforms these sub-goals to corresponding plans. Later, each plan is mapped to the equivalent BPEL4WS plan describing the composite Web Service. In [13], the user provides context information manually, however in our work, we use context sources and context agents to automatically obtain context information as well as changes to the context of the service user. Thus, we provide more concrete support for the acquisition of context information.

Besides the use of context information for service composition, there has been work reported on composing higher-level context from lower-level context elements. [12] proposes to use compositions of basic context elements to build higher-level contexts. A context composition mechanism, upon receipt of a request for higher-level context, forms all equivalent context expressions from the lower-level context elements and determines which equivalent context expressions can be instantiated. The CASE platform is targeted to achieve composition of all the services and therefore, can also be used to compose higher-level context using the context information gathered from lower level context sources.

The CASE platform dynamically re-composes services by subscribing to and utilizing changes in context information of both clients and services. This mechanism promises to simplify the design of clients in pervasive environments as they need not explicitly search and compose the best services when their (or the currently composed services) context changes. The other main aspect of our work is that the context sources hosted on mobile devices are modelled as services which can participate in service discovery to offer context information (and changes to such information) in a standardized way. The use of ontologies for the representation of context information ensures that services, context sources, and the CASE discovery and composition services can meaningfully share context information. However, it is further possible to improve the matchmaking behaviour of the CASE composition service by using the COIF function of [11].

6 Summary

Future 4G service platforms will need to be able to dynamically (re)compose services to deal with the frequent context changes inherent to 4G systems and the pervasive computing paradigm in general. Service composition is also required since statically-composed services will often not directly match requests for specific non-trivial services.

Dynamic service composition relies heavily on service discovery. The use of context information during discovery reduces the number of candidate services that match the discovery request which is essential in pervasive environments, where there may be many discoverable services.

The combination of context-aware service discovery and dynamic composition can be considered to be the "next level" of intelligent service discovery/matchmaking. We expect that in future 4G service platforms there will be an increasing need for such additional intelligence to aid the development of genuinely adaptive end-user applications. CASE is an example of a set of services that will be part of these platforms and is a step in the direction of more intelligent pervasive computing service platforms.

Acknowledgments. The authors would like to thank colleagues in the IST Amigo, Freeband Awareness, and Freeband AMUSE projects who have contributed to the work described in this article.

References

1. "AWARENESS Service Infrastructure D2.10 - Architectural specification of the service infrastructure", https://doc.telin.nl/dscgi/ds.py/ViewProps/File-47455
2. O. Ratsimor, V. Korolev, A. Joshi, and T. Finin, "Agents2Go: An Infrastructure for Location-Dependent Service Discovery in the Mobile Electronic Commerce Environment", ACM Mobile Commerce Workshop, July 2001
3. M. Etoh, "Beyond 3G: From3G To Seamless Intertechnology Wireless Networks", http://www.docomolabs-usa.com/pdf/PS2003-062.pdf

4. "OWL-S: Semantic Markup for Web Services", http://www.daml.org/services/owl-s/1.0/owl-s.html

5. K. Fujii and T. Suda, "Dynamic Service Composition Using Semantic Information", 2nd ACM International Conference on Service Oriented Computing (ICSOC '04), November 2004

6. V. Oleshchuk and A. Pedersen, "Ontology Based Semantic Similarity Comparison of Documents", 14th International Workshop on Database and Expert Systems Applications (DEXA'03)

7. F. Zhu, M. Mutka, and L. Ni, "Service Discovery in Pervasive Computing Environments", IEEE Pervasive Computing, October-December 2005

8. C. Noda, et al., "Distributed Middleware for User Centric System", 9th WWRF, Zurich, Switzerland, July 2003

9. HP Labs, "Jena – A Semantic Web Framework for Java", http://jena.sourceforge.net/, October 2005

10. Sun Microsystems, "Reggie: Sun Microsystems Jini Lookup service implementation", Jini Technology Starter Kit v2.1, http://starterkit.jini.org/downloads/index.html, October 2005

11. S. K. Mostefaoui, A. Tafat-Bouzid, and B. Hirsbrunner. "Using Context Information for Service Discovery and Composition." Proceedings of the Fifth International Conference on Information Integration and Web-based Applications and Services, iiWAS'03, Jakarta, Indonesia, 15 - 17 September 2003. pp. 129-138.

12. G. Thomson, S. Terzis, and P. Nixon, "Towards Dynamic Context Discovery and Composition", 1st UK-UbiNet Workshop, Imperial College, London, England, September 2003.

13. M. Vukovic and P. Robinson, "Adaptive, planning-based, Web service composition for context awareness", International Conference on Pervasive Computing, Vienna, April 2004.

14. Sun Microsystems, "The JINI Architecture Specification", http://www.sun.com/software/JINI/ specs/ JINI1_2.pdf, December 2001.

15. Sun Microsystems, "JINI Technology Surrogate Architecture Specification", http://surrogate.JINI.org/sa.pdf, October 2003.

16. Dey, D. Salber, and G. Abowd, "A Conceptual Framework and a Toolkit for Supporting the Rapid Prototyping of Context-Aware Applications", Special issue on context-aware computing in the Human-Computer Interaction (HCI) Journal, Volume 16 (2-4), 2001, pp. 97-166.

17. P. Debaty and D. Caswell, "Uniform Web presence architecture for people, places, and things", IEEE Personal Communications, Volume 8, Issue 4, Aug 2001, pp. 46-51

18. J. Rosenberg, H. Schulzrinne, G. Camarillo, A. Johnston, J. Peterson, R. Sparks, M. Handley, and E. Schooler, "SIP: Session Initiation Protocol", RFC 3261, June 2002

19. J. Kalaoja, J. Kantorovitch, S. Carro, J. María Miranda, Á. Ramos and J. Parra, "The Vocabulary Ontology Engineering", 8th International Conference on Enterprise Information Systems, Paphos, Cyprus, May 2006

Infrastructural Support for Dynamic Context Bindings

Tom Broens, Aart van Halteren, and Marten van Sinderen

Centre for Telematics and Information Technology
University of Twente, P.O. Box 217, 7500 AE Enschede, The Netherlands
{t.h.f.broens, a.t.vanhalteren, m.j.vansinderen}@utwente.nl

Abstract. Research in context-aware systems shows that using context information enables the development of personalized mobile applications. The context
acquisition process in a context-aware (CA) system consists of two main roles:
context producing entities (e.g. wrapped sensors) and context consuming entities (e.g. CA application). A CA system can be seen as a hierarchy of associated
context producers and consumers which exchange contextual information.
Managing contextual information used in context aware systems, introduces
additional complexity for mobile application developers. We focus on the dynamic processes of discovery, selection, (re)binding and monitoring of entities
that produce context information. Dynamic context binding is complex because
of the dynamic nature, in terms of availability and quality, of context producers.
We propose to delegate the responsibility for context binding to the middleware infrastructure and provide application designers with a declarative language to specify context information requirements on a high-level of abstraction. In this way, our
Context-Aware Component Infrastructure (CACI) provides support for dynamic
context bindings between application components and context producers.

1 Introduction

Ubiquitous computing envisions computer systems everywhere, which aid users in a
tailored and unobtrusive manner [1]. Context-awareness is a major enabler of this
paradigm and offers promising ways to adapt and personalize (mobile) applications.
Context-aware (CA) systems take besides explicit user input also the context of an
entity (i.e. person, place or object relevant to the functioning of the system) into account, to provide functionality which is adapted to the users situation [2]. This is
particularly interesting for mobile applications because these application function in
constantly changing environments due to the movement of the user [3]. For example,
a mobile tourist guide application could benefit from context by offering personalized
tourist information based on the current physical location of the user [4].

Main elements in CA systems are software entities that produce or consume
contextual information. Typical *context producers* are wrapped sensors that acquire
context from the physical environment (e.g. GPS, temperature sensor, ECG sensor).
Typical *context consumers* are CA applications that use context from context producers to adapt their functionality. Some software entities are both consumer and producer. For example context reasoners, which produce derived context information
based on context received from other context producers. A CA system can be seen as

P. Havinga et al. (Eds.): EUROSSC 2006, LNCS 4272, pp. 82–97, 2006.
© Springer-Verlag Berlin Heidelberg 2006

a hierarchy of associated context producers and consumers which exchange contextual information (see Figure 1). The association between a context consumer and a context producer is called a *context binding*. Exchanging contextual information consists then of two phases:

1. Create and maintain a context binding between a context consumer and producer (this includes discovery, selection and binding to a context producer by the context consumer).
2. Exchange of contextual information (this includes using a suitable context exchange protocol and context data format).

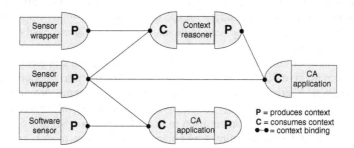

Fig. 1. Example hierarchy of context producers and consumers in a CA system

Due to missing infrastructural support, developers of first generation CA systems (e.g. [5]) programmed bindings between context consumers and context producers in an ad-hoc and tightly-coupled fashion, unique for a specific application [2, 6]. The developers choose specific context producers (e.g. GPS location sensors, RFID sensors) and program the low-level interaction between the specific context producer and his application (context consumer). Thereby, he creates a tight coupling between his application and the used context producers. Reuse of the created application is limited and future evolutions (e.g. upcoming of new technology) becomes difficult [2, 7].

Currently, there is trend towards middleware infrastructures for CA systems [8]. These infrastructures offer solutions to recurring problems in the CA domain, like context discovery, reasoning, adaptation and security. Using the run-time discovery and binding mechanisms these infrastructures offer, context producers and context consumers are decoupled and can be bound at run-time. However, establishing and maintaining the context binding is not trivial and still needs extensive programming effort [9]. Developers need to still create programming code to discover, select and bind to relevant context producers for every context consuming entity in the system. Furthermore, due to the mobility of the user, or possibly the context producer, the availability of the context producer for the context consumer is not guaranteed and reliable [10]. Maintaining the binding is therefore complex and needs additional programming effort to develop a flexible and robust context-aware system that can handle this dynamicity.

In this paper, we propose to shift the responsibilities of establishing and maintaining context bindings to the infrastructure and offer a generic binding transparency to application developers of CA systems. With our infrastructure (called CACI), we offer a binding specification language which enables developers to specify required bindings on a higher level of abstraction. Using this specification, application developers define the context information requirements in stead of specifying the specific context producer they need. Furthermore, CACI provides mechanisms to transform the binding specification from the developer into dynamic bindings which react to changes in availability of context producers. Infrastructural support for dynamic context bindings decreases programming effort for binding and the general complexity of CA applications. Furthermore, it provides a generic approach which is suitable for different types of CA system, thereby stimulating reuse. Additionally, it is a way to get a separation of concerns of the non-functional requirements of context binding and the functional requirements of the CA system. This leaves the application developer with more room to concentrate on the CA application logic. Summarizing, the goal of CACI is to facilitate application developers in quickly and easily developing a flexible and robust CA system.

The remainder of this paper is structured as follows: section 2 presents two real-life scenarios which give further motivation for an infrastructure supporting dynamic context bindings. Section 3 discusses an analysis of CA systems, providing the foundation for our infrastructure. Section 4 provides background information on component-based application development used in our infrastructure. Section 5 discusses the design of the CACI infrastructure, including CA component model, component description language and the CACI binding mechanisms. Section 6 presents related work and finally section 7 gives conclusions and directions for future work.

2 Scenarios

The following two scenarios indicate possible real-life applications of CA systems. The first scenario discusses a CA office environment while the second presents a CA healthcare environment.

2.1 CA Office Environment

Jerry works at a large bank. He is an account manager which is often preparing for meetings. He is looking for his colleague John, to give him a report for his next appointment. Jerry's CA system indicates that John's location is unknown. A moment later, John walks into the office building and his system logs into the CA office system. At that moment Jerry's system notifies him that John is walking towards the canteen. John can now intercept Jerry and transfer the document.

2.2 CA Healthcare Environment

Sophie is an epileptic patient suffering from regular seizures. Currently, she wears an epilepsy safety system (ESS) that enables her to be mobile with a feeling of safety. Sophie is walking to the groceries store and the ESS detects a likely occurrence of a seizure. Sophie is warned to sit down and relax while the health care center can notify a nearby and available care giver (e.g. family doctor, voluntary care giver, ambulance personal) of her state. The selected care giver can proceed to her location to provide first aid [11].

2.3 Discussion

Context-aware systems acquire and process contextual information. This requires a binding between context consuming applications and context producers. In the case of the office scenario, Jerry's system needs location of John. In the case of ESS, Sophie's and the healthcare system need the location of Sophie and the care givers, and the availability of care givers. Without a supporting infrastructure the developer of these applications needs to implement low-level interactions with specific context producers for every piece of context information. Current CA infrastructures support developers by offering discovery mechanisms and context exchange mechanism. However the dynamic properties of context bindings are neglected. For instance, in the office environment Jerry's system needs to get John's location (from a context producer that provides John's location) which is at that time not available. It is just not known which context producer will pop-up or leave at what time. To create a flexible and robust system, developers have to cope with this scenario and need to program some kind of complex monitoring strategy (when supporting functions are at all offered by the infrastructure) that incorporates all foreseen situations. Similarly in the healthcare environment, Sophie is mobile and moves between administrative domains that each exports all kind of different context producers with different properties (e.g. quality of context information). Based on the available context producers the right one has to be bound and this binding has to be monitored. For example, Sophie walks past an electronic shop that reads her RFID tag and can determine that she is in front of the electronic shop. This context producer provides Sophie's location to her. When she moves out of the domain of that store this context producer will disappear and another location producer has to be bound. Without a supporting infrastructure, developers need to program this complex binding strategy (discovery, selection, (re)binding and monitoring). We propose to relieve application developers from this task and to shift this responsibility to the infrastructure using high-level context binding specifications.

3 Analysis of Context-Aware Systems

Characteristics of context-aware systems and context producers influence the binding process. Therefore they need to be taken into account in our context binding mechanism.

3.1 Context and Context-Awareness

The most used definition of context is provided by Dey [2]. He defines *context* as any information that can be used to characterize the situation of an entity. An entity can be a human, a physical or a computational object. Common categorizations of context are: computing context (e.g. email received, bandwidth available), social context (e.g. health, mood, calendar) and physical context (e.g. location, temperature) [12]. *Context-awareness* is defined as a property of a system that uses context to provide relevant information and/or service to the user, where relevancy depends on the user's task.

Context consists of several layers of abstraction (see Figure 2, based on [13]). First, context is always related to an *entity*, which can be a person or an object. Without this entity information, context has no meaning [14]. Then context consists of the real contextual information which is made up out of the context *element* (e.g. location), *value* (e.g. 50.234, 6.152) and *format* (e.g. lat/long). Finally, context has meta properties like *security* policies (e.g. who may read this context) and *quality of context* (e.g. accuracy) [15].

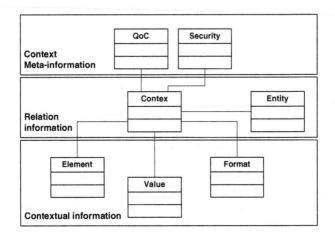

Fig. 2. Context model

3.2 Context Producer Properties

Context producers have some properties which indicate the main challenges our context binding transparency addresses:

- *Distributed:* context is offered by a multitude of physically distributed context producers. Problems that arise are how-to discover relevant context sources and how-to exchange context information. Current CA infrastructures focus mainly on this aspect. We position our infrastructure as an extension of current CA infrastructures and leverage on their facilities.

- *Dynamic availability:* the visibility of context producers for context consumers is subject to change. They can appear and disappear at arbitrary moment. For instance, when a user moves out a certain domain the context sources of that domain will disappear (see Figure 3). The availability of context producers is therefore not guaranteed and reliable.

- *Dynamic quality of context:* Furthermore, the quality of the produced context, i.e. Quality of Context (QoC) [15], can vary among context producers and also among context samples provided by a single context producer. Therefore, providing high quality CA applications requires incorporating dynamic quality aspects.

- *Heterogeneous context models:* (similar) context can be provided by different context producers using different data models for storing, accessing and transferring contextual information.

This paper focuses on overcoming the indicated dynamicity challenges (availability and QoC) of context producers. We consider the heterogeneity aspect as a future extension.

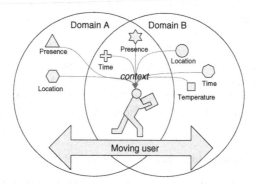

Fig. 3. Dynamic availability of context producers

4 Context-Aware Component-Based Application Development

A common type of infrastructure is middleware, which gives some advantages for the development of CA systems. CACI applies the component-based middleware paradigm.

4.1 Middleware and CA Systems

Middleware is often characterized as a software layer between applications and the underlying hardware platform. Main goals of middleware are [16]:

- Provide interoperability between distributed applications across heterogeneous platforms;
- Offer programming abstractions that hide complexities of building a distributed application. These abstractions are called transparencies [17] (e.g. the location transparency hides the physical location of the application instead of using a logical name or unique identifier to contact the distributed application).
- Offer common building block that solves recurring problems and thereby facilitate the development of distributed applications.

Middleware technologies have evolved, amongst others, from Procedural, Transactional, Message-oriented, Object-Oriented towards Component-based middleware. In general, we saw this evolution in middleware mechanisms to be able to manage the increasingly complexity of distributed applications. Context-awareness adds another layer of complexity to distributed applications (see section 3.2). To fully enable the ubiquitous computing vision, CA systems should provide a loose coupling between context producers and context consumer. A CA application (context consumer) should be able to use arbitrary context producers based on their availability and offered characteristics.

In general, middleware infrastructure approaches are beneficial for context-aware systems. If we consider the basic features of middleware infrastructures they facilitate the challenges that developing context-aware systems pose:

- Context producers and context consumers are distributed on possible hetero-geneous platforms. Middleware infrastructures can facilitate interoperability between these entities.
- As indicated, developing CA systems is complex. Therefore, there is a need for common abstractions that hide the complexity of specific aspects (e.g. binding, secure access) in the development of CA systems.
- Recurring problems like context discovery, security of context information, context-based adaptation and binding can be bundled into generic middle-ware building blocks.

4.2 Component-Based Middleware for CA Systems

Component-based middleware views an application as a composition of components. Szyperski [18] defines a *component* as: "a unit of composition with contractually specified interfaces and explicit context dependencies only. A software component can be deployed independently and is subject to composition by third parties". Figure 4 indicates the generic architecture of component-based middleware.

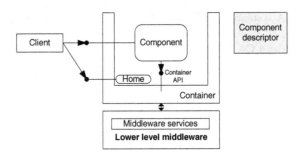

Fig. 4. Component-based middleware

Components are encapsulated by a container that offers the execution environment for components. They are deployed in the container using a component descriptor. Components interact with the container using the container API. The container (and also components) interacts with clients using some type of communication mecha-nism offered by the lower-level middleware. Generally, lower-level middleware infrastructures offer generic middleware services like transaction support, security, persistency and notification. The component offers certain functionality specified in interfaces to clients. The client can instantiate, destroy and find components using the home interface. The client receives a reference (or handle) to the component on which it can invoke remote method calls. Examples of currently available component-based middleware technologies are Corba Components, J2ME, J2EE and OSGi.

Traditional non-component-based applications can be modeled as a function that transforms an explicit user input into an output. With CA this model is extended with

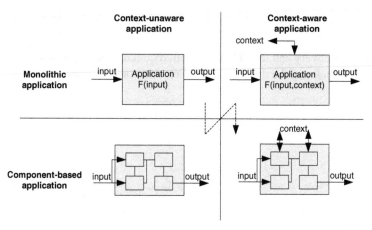

Fig. 5. From monolithic context-unaware applications to component-based CA applications

a context input offered by some context producer. If we then zoom into an application and apply the component-based paradigm, a CA application becomes a composition of CA components. The component can have individual context requirement and context offerings (see figure 5).

Current context-aware infrastructures are mainly OO-based solutions. We claim that a direction towards component-based middleware approaches is beneficial for the easy development of context-aware applications because:

- *Reusability of components:* this is a common advantage of components. They are well-defined encapsulated units of programming that can be easily reused.
- *Third party composition:* When developing a CA system, context consumers and context producers are subject to third party composition. Components are well suited for this third party composition.
- *Unit of deployment:* components execute in a run-time environment (often called container) this means that on deploy-time this environment can execute certain functionality. For CA systems this could includes initializing the context bindings and setting security policies.

5   CACI Architecture and Design

CACI offers a component-based approach to context-aware applications (see Figure 6). It considers a context-aware application as a composition of context-aware components. These components are deployed in a component environment called the container. The boundaries of a component are described using application ports and context ports. Application ports (like in CCM [19]) are descriptions of the functionalities a component requires and can offer. Generally, this is the functional interface of a component. In CACI, we take a similar approach for context ports. Context ports define the context requirements or offerings a component has. These ports are described in the component descriptor which is used at deploy-time to configure the container. Configuration actions the CACI binding mechanism takes based on the descriptor are:

- Discovery and initial binding with required context producers.
- Initializing the dynamic re-binding monitoring process.

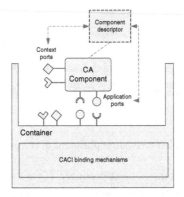

Fig. 6. CACI component model

The remainder of this section discusses CACI in further detail. It starts with a description of the CACI component description language. Then, it presents details on the design of CACI binding mechanism. Finally, it discusses the internal design of a CACI CA component.

5.1 CACI Component Description Language

The CACI Component Description Language (CCDL) describes the context requirements of a CACI component. The description is the knowledge base for deployment and operational management of this component and its binding. Figure 7 presents the meta-model of CCDL expressed in UML.

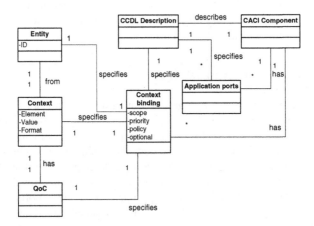

Fig. 7. Meta-model CCDL

CCDL describes, besides 'standard' application ports, context binding specifications (i.e. context ports) relevant to the component. A context binding specifies which context is required or offered. Currently we only consider context requirements, facilitating the development of context producers by offering a context offering specification is considered future work. Furthermore, the specification specifies to which entity this context belongs and which quality parameters it has or should have. Furthermore, it specifies the *scope* of the binding. Scope describes if the binding can only be to context sources within the container (local) or also from the environment (global).

Additional parameters for the context bindings are:

- *Priority:* indicates the order to resolve the binding.
- *Policy:* can be dynamic, semi-dynamic or static. Dynamic indicates that the binding has to be monitored at run-time and when 'better' context producers pop-up, a re-binding process has to start. Additionally, when the already bound context producers leave, a new context producer has to be bound. Semi-dynamic indicates that only when an already bound context producer leaves, a new context producer has to be bound. Static indicates that binding will be done once at deploy-time.
- *Optional:* is a Boolean value which indicates if the deployment of the component should fail when a binding cannot be resolved.

Figure 8 gives examples of CCDL binding specifications in pseudo-code, useful for the two scenario's sketched in section 2. CCDL is still under research. The information model, for instance how-to define QoC criteria, is subject of future research.

```
CA_office_env_context_binding ::
  requires Location
  from Colleague.John
  expressed_in LatLong
  with accuracy > 75%
  scope global
  policy semi-static

CA_healthcare_env_context_binding ::
  requires Availability
  from Doctor.Smith
  expressed_in Boolean
  with accuracy > 95%
  scope global
  policy dynamic
```

Fig. 8. Example of a CCDL context binding specification

5.2 CACI Binding Mechanism

Figure 9, gives an overview of the deployment and operational phase of a CACI CA component. When a component is deployed, its CCDL descriptor is parsed by the deployer. The deployer sends a binding request to the context binder which generates and instantiates a context producer proxy. It discovers other local container context sources and when appropriate (scope is global), it sends a request to the underlying infrastructure for discovery of relevant context producers. Based on the description it binds the generated proxy to a remote context producer or, when none can be found, informs the component the binding cannot be resolved.

Besides this initial binding process, it generates context binding tracking rules (determined by the policy specified in the binding specification) which are used to monitor the binding and possibly initiate re-binding.

Using a context producer proxy has the following advantages:

- *Transparency:* by shielding the component for direct interaction with the remote context producers these components become unaware of dynamicity of remote context producers. The local proxy act as a homogenizing layer.
- *Optimization:* the proxy enables to have a different acquisition strategy between the CA component and the proxy and the proxy and the remote context producer. This acquisition strategy can be optimized dynamically based on the context producer or underlying network technology (e.g. buffering, caching).

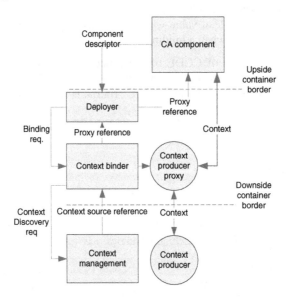

Fig. 9. Context producing component

The context binder applies the dynamic context binding algorithm. Figure 10 shows a high-level activity diagram of this algorithm. When a component is deployed its binding specification is interpreted and a binding is tried to establish (unbound state). When this process is a failure the component is either not deployed or the binding is ignored (based on the optional statement in the component description). When there is successful binding (bound state) a re-binding process is initiated based on the decision by the context binder. This decision making algorithm is based on the

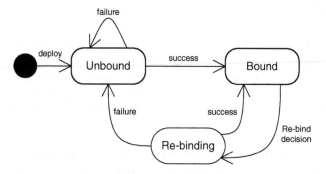

Fig. 10. Binding algorithm

disappearing of the currently bound context producer or appearing better context producers. This decision making algorithm is still currently subject of research.

5.3 CACI Internal Design

A major goal of CACI is to offer a binding transparency which has the features of ODP location and relocation transparencies [17]. To offer these transparencies, CACI applies the proxy pattern [20] for its internal representation of a deployed CACI CA component (see Figure 11). The proxy pattern shields the CACI context consuming component from the remote context producer. It creates an intermediary proxy component. This proxy is a representative of the context producing component and performs processing needed for maintaining the binding.

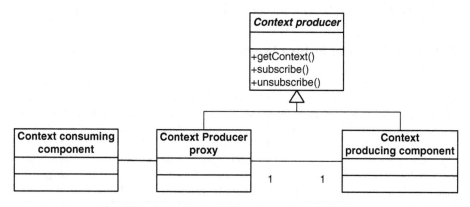

Fig. 11. Appliance of the proxy pattern in CACI

We support two common types of context acquisition strategies:

- *Request-Response:* the context consuming component explicitly requests for a context sample from the context producer.
- *Subscribe-Notify:* the context consuming component subscribes to context updates triggered by the context producer.

6 Related Work

Several context-aware middleware infrastructures [8, 21-23] offer context management functionality. Generally, this functionality enables context consumers to discover and bind to context producers using programming statements. Often dynamic
monitoring capabilities (e.g. notification when new and possibly better, context
sources arise) are not available and when context producers become (un)available the
decision to re-binding and the choice to bind to which context source has to be taken
by the application rather than the infrastructure.

Cervantas [24] proposes a component model that enables autonomous adaptation
of the binding of component services at run-time, using a component description
language. These principles are implemented in OSGi as the Service binder. Furthermore, this service is the foundation for the OSGi Declarative service, adopted in
release 4 of the OSGi specification [25, 26]. The service binder solely considers components, which may require or offer services, within the scope of one OSGi container,
thereby limiting the scope of binding to a singe computing system. Furthermore, the
scope of the service binder is generic services. Therefore the capabilities of the specification language and its corresponding mechanisms are not tailored to context specific
needs. This implies that specific aspects of context like QoC are not supported. QoC
highly influence the binding process and therefore needs a central place. However, we
adopted some of the aspects supported in their component description language.

Bottaro [27] extends the Service binder principles towards distributed services.
Furthermore, dynamic re-binding based on service availability is supported. The binding choice is not specified in the component specification but has to be programmed.
Context information can be retrieved and stored (on registration) as properties which
can be used by the binding decision. However, dynamic context quality changes that
trigger rebinding, throughout the lifetime of the service, are not incorporated.

7 Conclusions and Future Work

Context is important for ubiquitous environments that want to offer personalized
applications in an unobtrusive way. Context producers are the providers of context
information. Binding to context producer is challenging because of their distributed,
heterogeneous and dynamic nature. These context sources offer the information for
context-aware applications to adapt their behavior.

In this paper we introduce CACI which is an application infrastructure simplifying
the development of context-aware applications. CACI is based on the component-
based middleware paradigm which generally improves reusability and flexibility
compared to traditional OO- and RPC-based middleware paradigms. Specifically,
CACI introduces a context-aware component model which defines that a component
besides application ports also consists of context ports. These context ports specify
the context requirements and the context offerings of the component. Especially the
context requirements can be influenced by context meta-properties like QoC.

Further, CACI offers a component description language that can be used to de-
claratively specify a component's context requirements and context offerings. This
specification is used by the CACI infrastructure to resolve at run-time the context

requirements or registering the context offerings to the global environment, depending on the component's role as context consumer or context producer, respectively. This context description is a natural extension of already available component descriptions. It decreases programming code and the general complexity of the component. It is also a way to get a separation of concerns of the non-functional requirements of context binding and the functional requirements of the component.

With CACI, the binding between context producing and context consuming components is flexible. When context producers/consumers leave or enter the system the bindings are re-evaluated and when suitable a seamless rebinding process is started.

We implemented a proof-of-concept prototype based on the OSGi component framework [28] using the Oscar implementation [29]. The OSGi framework offers lifecycle management of components (called bundles). This includes installation, starting, stopping and updating of components. We reuse the powerful life-cycle management facilities of OSGi and extend it with CACI's context-aware component mechanisms. This extension consists of deploying the CACI binding functionality as a standard component in the container. A CACI component adds a pointer to a CCDL description in its component descriptor (i.e. manifest in the jar file of bundle). At deploy time the context manager parses the CCDL description and establishes bindings. At runtime these binding are monitored. When context sources pop-up or leave, the bindings are evaluated and when necessary a re-binding process is initiated. CACI runs on mobile or fixed system capable of running a java J2ME PP1.0 virtual machine (i.e. in case of windows mobile, IBM J9 is tested). Although CACI is currently implemented on top of OSGi, its principles are not specific for this technology and will be validated in the future on other platforms.

In this paper we present our efforts toward an infrastructure supporting the easy and rapid development of CA applications. However several aspects remain for future research:

- *Support for developing context producers:* currently CACI focuses on easily establishing context bindings for context consumers. However similarly, we could support the developers of context producers by letting them specify their context offerings in CCDL. CACI can then do the registration to the environment and tackle issues like access control to a producer and graceful degradation of a context producer in case of a malfunctioning sensor.

- *Binding algorithm:* further research is needed in the monitoring and rebinding mechanism of CACI. Questions that arise are how-to sense a lost binding, which policy is needed to make a decision on rebinding and how-to do this seamless re-binding.

- *Formalization of CCDL:* the basic structure of CCDL is presented in this paper. However, the information model supporting CCDL needs further research.

- *Stability:* dynamic re-binding has advantages but could also influence the stability of the system. Research is needed on how-to avoid oscillating bindings between context consumers and producer.

- *Proof-of-concept:* To indicate the generality of the CACI approach we plan to deploy CACI on other lower-level middleware technologies like the Jini Service Container [30].

Acknowledgement

This work is part of the Freeband AWARENESS Project. Freeband is sponsored by the Dutch government under contract BSIK 03025. (http://awareness.freeband.nl).

References

1. Weiser, M. and J. Brown, *The Coming Age of Calm Technology.* 1996.
2. Dey, A., *Providing Architectural Support for Context-Aware applications.* 2000, Georgia Institute of Technology.
3. Satyanarayanan, M., *Fundamental Challenges in Mobile Computing*, in *Fifteenth ACM Symposium on Principles of Distributed Computing*: Philadelphia, USA.
4. Pokraev, S., et al., *Service Platform for Rapid Development and Deployment of Context-Aware, Mobile Applications*, in *International Conference on Webservices (ICWS'05), Industry track.* 2005: Orlando, Florida, USA.
5. Korkea-aho, M., *Context-Aware Applications Survey*, 2000, Available from: http://users.tkk.fi/~mkorkeaa/doc/context-aware.html.
6. Pascoe, J., *Context-aware software.* 2001, University of Kent: Canterbury.
7. Ebling, M., et al., *Issues for Context Services for Pervasive Computing*, in *Workshop on Middleware for Mobile Computing.* 2001: Heidelberg, Germany.
8. Henricksen, K., et al., *Middleware for Distributed Context-Aware Systems*, in *DOA 2005.* 2005, Springer Verlag: Agia Napa, Cyprus.
9. Banavar, G. and A. Bernstein, *Software infrastructure and design challenges for ubiquitous computing applications.* Communications of the ACM, 2002. **45**(12): p. 92-96.
10. Bellavista, P., et al., *Dynamic Binding in Mobile Applications.* IEEE Internet Computing, 2003. **March-April**: p. 34-42.
11. Broens, T., et al., *Towards an application framework for context-aware m-health applications*, in *EUNICE: Networked Applications (EUNICE'05).* 2005, ISBN: 84-89315-43-4: Madrid, Spain.
12. Schilit, B., N. Adams, and R. Want, *Context-Aware Computing Applications*, in *IEEE Workshop on Mobile Computing Systems and Applications.* 1994: Santa Cruz, CA, USA.
13. Broens, T. and A. Halteren, *SimuContext: Simply Simulate Context*, in *International Conference on Autonomic and Autonomous Systems (ICAS'06).* 2006: Silicon Valley, USA.
14. Dockhorn Costa, P., et al., *Towards Conceptual Foundations for Context-Aware Applications*, in *Third International Workshop on Modeling and Retrieval of Context (MRC'06).* 2006: Boston, USA.
15. Bucholz, T., A. Kupper, and M. Schiffers, *Quality of Context: What It Is And Why We Need It*, in *Workshop of the HP OpenView University Association 2003 (HPOVUA 2003).* 2003: Geneva.
16. Emmerich, W., *Software Engineering and Middleware: A Roadmap*, in *22th ICSE 2000.* 2000: Limerick, Ireland.
17. Blair, G. and J. Stefani, *Open Distributed Processing and Multimedia.* 1998: Addison-Wesley.
18. Szyperski, C., D. Gruntz, and S. Murer, *Component Software: Beyond Object-Oriented Programming.* Component Software serie, ed. C. Szyperski. 2002: Addison-Wesley.
19. Wang, N., D. Schmidt, and O. O'Ryan, *Overview of the CORBA Component Model*, in *Component-based software engineering: putting the pieces together.* 2001. p. 557-571.

20. Buschmann, F., et al., *Patter-oriented software architecture: a system of patterns*. 1996: Wiley.
21. Kranenburg, H. and H. Eertink, *Processing Heterogeneous Context Information*, in *Next Generation IP-based Service Platforms for Future Mobile Systems workshop*. 2005: Trento, Italy.
22. Kummerfeld, B., et al., *Merino:Towards an intelligent environment architecture for multi-granularity context description*, in *workshop on User Modelling for Ubiquitous Computing*. 2003.
23. Bardram, J., *The Java Context Awareness Framework (JCAF) - A Service Infrastructure and Programming Framework for Context-Aware Applications*, in *Pervasive Computing*. 2005: Munchen, Germany.
24. Cervantas, H. and R. Hall, *Autonomous Adaptation to Dynamic Availability Using a Service-Oriented Component Model*, in *26st International Conference on Software Engineering*. 2004: Edinburgh, Scotland.
25. OSGi Alliance, *OSGi Service Platform Core Specification: Release 4*. 2005.
26. OSGi Alliance, *OSGi Service Platform Service Compendium: Release 4*, 2005.
27. Bottaro, A. and A. Gerodolle, *Extended Service Binder: Dynamic Service Availability Management in Ambient Intelligence*, in *International Workshop on Future Research Challenges for Software and Services (FRCSS'06)*. 2006: Vienna, Austria.
28. OSGi Alliance, *The OSGi Service Platform - Dynamic services for networked devices*, 2005, Available from: http://osgi.org.
29. Oscar.org, *Oscar - An OSGi framework implementation*, 2005, Available from: http://oscar.objectweb.org/.
30. Cheiron, *Cheiron project site*, Visited 2006, Available from: http://www.cheiron.org/.

Adding Context Awareness to C#

Anca Rarau and Ioan Salomie

Computer Science Department
Technical University of Cluj-Napoca, Romania
{Anca.Rarau, Ioan.Salomie}@cs.utcluj.ro

Abstract. In this paper, we propose a context-aware extension of the
C# programming language. Bringing the context-awareness at the level
of the programming language constructs comes with the advantage of
lowering the effort required to specify the contextual polymorphic behavior of the applications. The application developer is released, at a great
extent, from the responsibility of managing the context-awareness without reducing the programmer's ability to express the context-dependent
behavior. In order to insert context-awareness into the C# programming
language we have extended the programming language constructs (i.e.
constant, variable, function, object) with context awareness. We have
also provided language constructs for dealing with the context model.

1 Introduction

Mark Weiser [14] was presenting a vision upon a world in which technology is
spread all over around us and is serving us in a transparent manner without
calling for our attention or calling for it to a minimum extent. This technology
is called ubiquitous or pervasive technology. There have been various research
trends that deal with the pervasive systems. One of these trends would be the
one that focuses on the context-aware systems. An application provided with
awareness to the context means an application capable of adapting itself to the
context in which it runs. Through this adaptation ability the interaction between the user and the computing device becomes freer, that is the amount of
necessary interaction and the attention required from the user decrease considerably. A context-aware application [12] is capable of extracting the information
and of structuring the data from the context in which it runs, thus sparing the
user from the explicit introduction of the input data. The extracted information
drives the behavior of the application. The application context may be given
by the following attributes: location (office, corridor, meeting room, etc), time
(hour, day, etc), activity performed by the user (walk, sleep, read, etc), nearby
resources (restaurant, exhibition, etc), nearby people (friend, boss, etc), technical parameters (available memory, bandwidth, etc), environmental parameters
(light, noise, humidity, temperature, etc), etc. The advantage of the context-
awareness is apparent when the applications are executed on the mobile devices
(e.g. PDA, mobile phone). Often the surroundings for these devices (and therefore the contexts in which their applications run) change quite fast as the people

P. Havinga et al. (Eds.): EUROSSC 2006, LNCS 4272, pp. 98–112, 2006.

use them while on the move. The interaction between the user and the application becomes more difficult as the user spreads the attention around and as the available I/O peripherals (e.g. screen, keyboard, etc) are rather limited. The developing of the context-aware applications implies a broad range of aspects to be regarded:

- **gathering of the context information**: smart sensors, sensor networks, positioning algorithms, etc;
- **management of the context information**: how to interpret and aggregate the context information, sensors fusion, context information quality improvement, how to make use not only of the current context but also of the past contexts, context forecast, etc;
- **application reaction to the context changes**: reaction algorithms, etc.

As far as the application reaction to the context changes is concerned, a couple of phases can be distinguished. At the beginning, the context-aware behavior was spread throughout the entire applications mainly being described by *if context then action* approach. This led to the development of hard to be maintained context-aware applications. In order to avoid this shortcoming, the context-aware behavior was organized into modules. As a result, the programmers were provided with the development tools such as libraries, frameworks and toolkits that facilitate the coding and the maintenance of the applications. Even when the development of the context-aware applications relies on these tools, the programmer still has great responsibility in managing the context-awareness of the applications (e.g. the programmer must be aware of the support infrastructure). Often this could be error-prone. Therefore, we have looked for new ways to handle the context-awareness so that to reduce the programmer's involvement in managing the context-awareness, without reducing the programmer's ability to express the context-dependent behavior. Thus we have brought the context-awareness at the level of the programming language constructs and we have proposed a context-aware extension of the C# programming language.

The rest of the paper is structured as follows. In section 2 we present the basic concepts of the context-aware extension of the C#. In section 3 we discuss the multifacet programming language constructs. In section 4, the interaction among multifacet programming constructs is presented. In order to validate our proposal we have both developed applications that use AwareC# (in order to verify AwareC# usability) and carried out the performance analysis (in order to verify the delays introduced by AwareC#). The validation of our proposal is discussed in section 5. In section 6 we present the related work while section 7 concludes the paper.

2   AwareC# Basic Concepts

Our approach to insert context-awareness into the C# programming language is divided into two steps. Firstly, we have added a context model at the language level. Secondly, we have extended the programming language constructs (i.e. constant, variable, function, object) with context awareness. In order to accomplish the two steps we have introduced the following notions as first class citizens into the C# language: *context* and *facet*. We have also extended the C# syntax with the following keywords:

- **context**, **template**: used to specify the contexts;
- **facet**: used to define the facets;
- **multifacet**: specifies that a programming language construct is context aware;
- **strict**: used when a context-aware object is declared. This keyword shows that the object is a strong typed object;
- **loose**: used when a context-aware object is declared. This keyword shows that the object is a weak typed object;
- **acquire**: used for facet acquiring operation;
- **discard**: used for facet discard operation.

2.1 Context

So far there has not been any kind of standard for context modeling. In [1], Chen and Kotz made a survey of context models used by different research groups. The models reviewed were: key-value pairs, tagged encoding (which leads to a more general approach based on XML), object-oriented model, logic-based model and the layered structure of the context. Apart from this context models, in another survey [8], the contextual graph and web-based model are discussed. There have also been other models for context modeling. Many of them are based on ontology. In [3], OWL and *Predicate(subject, value)* pattern are used to describe the context. OWL is also used in Cont-el [13]. In [6] ontology is employed for modeling the description of the context information from the sensors. Taking another approach, in [15], they proposed the following context taxonomy: device-specific context (e.g. number of objects running on the laptop, battery, etc), environment-specific context (e.g. current location, nearby devices, light level, temperature, etc) and user-specific context (e.g. personal profile, how many time a user launches an application, etc). In [9] they promote the 5W1H (who, where, what, when, why and how) context model. In [11] the context model is expressed as 4 parameter-predicate:

Context(<ContextType>,<Subject>,<Relater>,<Objects>).

Our approach regarding the context is somehow similar to [3] and [11] in the sense that we also use tuples to describe the context. In contrast with the approach that has been reviewed we directly include the context model within the programming language level. To our knowledge, there has not been much research regarding the encapsulation of the context model at the language level. In our model, the basic unit for the context description is the *elementary feature*. An elementary feature describes a property of an entity within the environment and comprises a type, the name of the entity, the name of property and the value of property (stated as the condition met by the property). A relational operator and a value give the condition. The name of entity can be left empty if the feature refers to the whole context. Two or more elementary features can be aggregated into a *compound feature*. Thus a compound feature describes one or more properties of one or more entities within the environment. Both elementary and compound features can be seen as sets (the elementary feature is a set made up of one item) so the operation valid for features got inspired from the set theory.

Operations upon features

1. Union (denoted by +): the result is a compound feature that contains all the elements of the operands;
2. Intersection (denoted by (&)): the result is a compound feature that contains all the elementary features common to the operands;
3. Difference (denoted by -): the result is a compound feature that contains those elementary features that belong to the first operand but do not belong to the second operand;
4. Comparison (denoted by == for equality and != for inequality):
   (a) two elementary features are equal if their types, the entities names, the properties names and the conditions are exactly the same;
   (b) two compound features are equal if they comprise the same number of elementary features and for each elementary feature from an operand there is exactly one equal elementary feature from the other operand.

The features can be aggregated into context structures (for short contexts). A context describes a state of the surroundings.

Definition. At a given moment, a context structure C is called current context if the state of the surroundings described by C matches the current state of the surroundings.

Simultaneously, more contexts can be current.

Operations upon context structures

1. Conjunction (denoted by &&): the result is a context structure that is current whenever all the operands are current;
2. Disjunction (denoted by ||): the result is a context structure that is current whenever one of the operands is current;
3. Negation (denoted by !): the result is a context structure that is current whenever the operand is not current;
4. Comparison (denoted by == for equality and != for inequality): two context structures are equal if whenever one operand is current the other operand is current too.

2.2 Facet

Our approach for inserting context-awareness into the C# programming language involves the ability to extend the programming language constructs (i.e. constant, variable, function and object) with context awareness. The facet is the mechanism that allows the programmers to handle the context awareness of the programming language constructs. One or more facets can be added / deleted to / from a constant, variable, function or object. A facet comprises a name, a context structure and an element (which can be a primitive value, a function body or an object). We say that the element is linked to the facet.

```
facet AFacet = AContext : AnElement;
```

The facet receives its type from the type of its element. E.g. a facet has float type if its element has float type. A facet, whose element is a function, has the type given by he signature of this function.

A facet can successfully be associated to a programming language construct if its type is compatible with the type of programming language construct. Whenever the context associated to a facet is the current context we say that the facet is in active state. A programming language construct can hold any number of facets. Anyway, the programming language construct must have a deterministic value / behavior, in the sense that, in a given context the programming language construct may have exactly one value (for constants and variables and objects) or one behavior (for functions and objects). Therefore, a programming language construct cannot have two or more facets having the same associated context. This would lead to some non-deterministic values / behaviors. Sometimes more than one facet associated to a programming language construct can be active simultaneously. This also leads to non-deterministic values / behaviors. In such a case, only one active facet is chosen to be exposed (the others are hide). The exposed facet will dictate the value / behavior of the programming language construct. In any moment the oldest active facet is exposed. If there is no facet exposed than the value / behavior of the programming language construct is given by an implicit face.

Operation upon facet

1. Comparison (denoted by == for equality and != for inequality): two facets are equal if both their contexts and their elements are equal.

3   Multifacet Programming Language Constructs

A programming construct with one or more facets it is called *multifacet programming construct*. Thus we get multifacet constants, multifacet variables, multifacet functions and multifacet objects.

3.1   Multifacet Constant

Definition. A multifacet constant is a constant that has one default value and an immutable value for each context it is sensitive to.

The values are independent to each other and all have the same type, which is the type of the multifacet constant. While the value of a common (not context-aware) constant cannot be changed during the program execution, the value of a multifacet constant can be changed when the context changes. But, given a context the value of the multifacet constant stays unchanged. If the current context is different than the contexts the multifacet constant is sensitive to, then the constant will expose the default value. The operations valid for a multifacet constant having a certain type are those valid for its not context-aware counterpart.

3.2   Multifacet Variable

Definition. A multifacet variable is a variable that has one default value and a value for each context it is sensitive to.

The values are independent to each other and all have the same type, which is the type of the multifacet variable. The multifacet variable holds as many (changeable) values as the number of contexts it is sensitive to. Given a current context, the multifacet variable exposes the value associated to that context. If the current context is different than the contexts the multifacet variable is sensitive to, then the variable will expose the default value. The operations valid for a multifacet variable having a certain type are those valid for its not context-aware counterpart.

3.3   Multifacet Function

Definition. A multifacet function is a function that has one default body and a body for each context it is sensitive to.

Given a current context, the body defined for that context gives the behavior of the multifacet function. If the current context is different than the contexts the multifacet function it is sensitive to, then the default body gives the function behavior. Thus the behavior of a multifacet function is tuned not only by the actual parameters but by the current context also. We have introduced the automatically call of a multifacet function upon the context change: if the function is in progress when the context changes than its body stops the execution, than a new body is selected (corresponding to the new current context) and this body is launched.

3.4   Multifacet Object

Definition. A multifacet object is an object that has a default structure and behavior (default object) and one structure and behavior for each context it is sensitive to.

The multifacet object can be considered as a collection of objects made up of the default object and one object for each facet (i.e. context it is sensitive to). Given a current context, the structure and behavior of the multifacet object is given by the structure and behavior defined for that context. If the current context is different than the contexts the multifacet variable is sensitive to, then the object will expose the default structure and behavior. The interface of the multifacet object consists of the set of the default object methods. We have implemented two types of multifacet objects:

1. A strong typed object is a multifacet object for which the objects linked to the facets have the same type as the type of the default object.
2. A weak typed object is a multifacet object that do not meet the condition stated for the strong typed object i.e. no type checking is performed when a facet is added.

We have adopted the delegation and consultation mechanisms [5] [7] as the core for the binding algorithm (i.e. binding the name of method to the body of method). For the strong type object the algorithm is as follows: if the multifacet object is not sensitive to the current context, the method of the default object is called. Otherwise, the consultation mechanism is triggered i.e. the default object delegates the object associated to the facet defined for the current context to launch the method. For the weak type object the algorithm is as follows: if the multifacet object is not sensitive to the current context, the method of the default object is called. Otherwise, either the delegation or the consultation mechanism is triggered. The delegation is applied if the object associated to the facet defined for the current context does not have the method. Otherwise the consultation is applied.

3.5   Operations

The declaration of multifacet constants, variables, functions and objects require the use of the `multifacet` keyword to precede the declaration in C#. In order to discriminate the two types of multifacet objects their declarations contain `strict` and `loose` respectively.

```
multifacet const int CaConstant;
multifacet float CaVariable;
multifacet int CaFunction(int a, int b);
multifacet strict AClass CaStrictObject;
multifacet loose AClass CaLooseObject;
```

The programmer uses the multifacet constants, variables, functions and objects in the same manner their common, not context sensitive, counterparts are used.

Apart from the common (not context-aware) operations on the multifacet programming constructs, the context-awareness adjusting has been defined. During the program execution a multifacet programming construct can increase its context-awareness by getting new facets (*acquire operation*) or can decrease its context-awareness by deleting some of its old facets (*discard operation*).

```
caConstant acquire AFacet;
caConstant discard AFacet;
```

3.6   Example

In the following example we illustrate the usage of the multifacet language programming constructs. Firstly, we define two context, one describing the `weekend` and the other describing the `weekdays`.

```
context Monday = [STRING, "time", "day", =, "monday"];
context Tuesday = [STRING, "time", "day", =, "tuesday"];
context Wednesday = [STRING, "time", "day", =, "wednesday"];
context Thursday = [STRING, "time", "day", =, "thursday"];
context Friday = [STRING, "time", "day", =, "friday"];
context Saturday = [STRING, "time", "day", =, "saturday"];
context Sunday = [STRING, "day", =, "sunday"];
context Weekdays = Monday || Tuesday || Wednesday || Thursday || Friday;
context Weekend = Saturday || Sunday;
```

Secondly, we create a multifacet constant `CaParkingTime` that holds the free parking time: in any weekday the free parking time is 15 minutes while during weekend it is 30 minutes. As `CaParkingTime` must hold two values, we have first to create two facets.

```
facet WeekdayParkingTime = Weekday : 15;
facet WeekendParkingTime = Weekend : 30;
```

Once the facets have been created, the multifacet constant will acquire both of them.

```
//multifacet constant declaration
multifacet int CaParkingTime;
CaParkingTime acquire WeekdayParkingTime;
CaParkingTime acquire WeekendParkingTime;
```

Then the value of a common (not context-aware) variable that keeps the current parking time of a car can be compared against the value of the multifacet constant. The value of `ParkingTime` will be compared against either 15 or 30 depending on the context.

```
if(ParkingTime > CaParkingTime)...
```

4   Interaction Among Multifacet Programming Constructs

In this section, we analyze what happens when two ore more multifacet programming constructs interact with each other. By interaction we mean that one multifacet construct uses (calls) other multifacet constructs. The issue is to have all the programming constructs, involved in the interaction, exposing facets corresponding to the same context.

Definition. The interaction between two ore more multifacet programming constructs is consistent if the facets exposed by the programming constructs during whole interaction correspond to the same context.

As an example, we consider the statement $a = b+c$. In this statement the multifacet variables a and b respectively a and c interact with each other in the sense that a uses the values of b and c. What happens if while the *add* and *assignment* operations are executed the multifacet variables a, b and c expose facets corresponding to different contexts? We take into consideration the scenario illustrated in figure 1. The multifacet variable b has three facets. Two of them are relevant for the scenario; the one corresponding to the C_1 context holds value 20 while the one corresponding to the C_2 context holds value 2. The c variable has also three facets. Again only two of them are relevant for the scenario illustrated in figure 1. For c the values corresponding to the C_1 and C_2 contexts are 2 and 1 respectively. In case the C_1 is the current context the expected outcome of the statement $a = b+c$ execution consists in assigning the value 22 to the a variable on the facet corresponding to the C_1. Let us consider the situation where the statement $a = b+c$ is executed while the current context is being changed from C_1 to C_2. Two actions are performed simultaneously. One action concerns the reaction of the multifacet variables a, b and c to the context changes i.e. the updating of the currently exposed facets of a, b and c. The other action deals with the actual execution of the statement. Each action goes on a separately execution thread. We call F_{MC} the thread that copes with the context change while we call F_{EI} the thread that concern the statement execution. The two simultaneous threads got interleaved. This may lead to wrong outcomes of the statement execution. Such a situation happens when the execution thread interleaving occurs as follows (where T_1, T_2, T_3, T_4 are ordered moments of times $T_1 < T_2 < T_3 < T_4$):

1. at T_1 on F_{MC}: hide the old current exposed facet of a variable (facet corresponding to C_1) and expose the new current exposed facet of a (the one corresponding to C_2);
2. at T_2 on F_{MC}: hide the old current exposed facet of b variable (facet corresponding to C_1) and expose the new current exposed facet of b (the one corresponding to C_2). The new current facet has 2 as a value.
3. at T_3 on F_{EI}: execute the $a = b+c$ statement whose result will be 4. This value goes to the exposed facet of a (the one corresponding to C_2).
4. at T_4 on F_{MC}: hide the old current exposed facet of c variable (facet corresponding to C_1) and expose the new current exposed facet of c (the one corresponding to C_2).

The above thread interleaving leads to an incorrect result, as the value computed for a in C_2 context is 4 instead of 3. The fact that the c variable has updated its exposed facet before the statement execution is the source of the incorrectness. The correct outcome would be obtained only if all the involved multifacet variables updated their exposed facet before statement execution.

In order to give to the application developer means of dealing with the consistent interaction, we have introduced *Synchronize* and *Release* primitives. The aim of the *Synchronize-Release* pair is to guarantee that a statement is executed only after all the involved multifacet programming constructs have reacted to the context changes. *Synchronize* implements a barrier mechanism. *Synchronize* ensures that the multifacet constructs provided as parameters become insensitive to the context changes. The *Synchronize* call forces all the multifacet constructs (provided as parameters) to react to the context change, if they have not

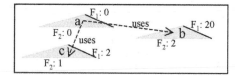

Initial state.

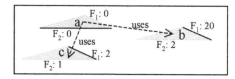

T_1: for a variable: the facet assigned to the C_1 context is hidden, while the facet assigned to C_2 context is exposed

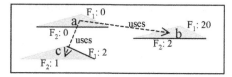

T_2 : for b variable: the facet assigned to the C_1 context is hidden, while the facet assigned to C_2 context is exposed

$$a = 2 + 2$$

T_3 : statement $a = b + c$ is executed

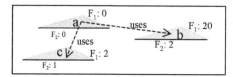

T_4 : for c variable: the facet assigned to the C_1 context is hidden, while the facet assigned to C_2 context is exposed

Fig. 1. Execution thread interleaving which leads to an incorrect results

reacted so far. Once they have reacted their sensitivity gets blocked until the *Release* primitive is called for them. *Release* not only gives back to the programming constructs the ability to react to the context changes, but it also forces them to react immediately to the current context thus updating their exposed facets.

In the current implementation, if many execution threads use the a multifacet variable, but one of them block the a's sensitivity then a is seen as being insensitive by all the other threads until a gets released.

If a variable have been blocked several times it reacquires its sensitivity only if it is released from all the previous *Synchronize* calls.

For the statement $a = b+c$ we can block the context sensitivity of a, b and c using the *Synchronize-Release* pair:

```
ManualResetEvent mse = Synchronize(a, b, c);
a = b + c;
Release(mes);
```

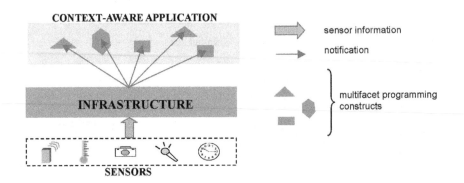

Fig. 2. The relationship between the environment, the infrastructure and the context-aware application

5   Validation

In order to validate our proposal we had both:

1. to develop applications that use AwareC# in order to prove AwareC# usability;
2. to perform the performance analysis in order to prove that the delays introduced by AwareC# are reasonable for a broad range of applications.

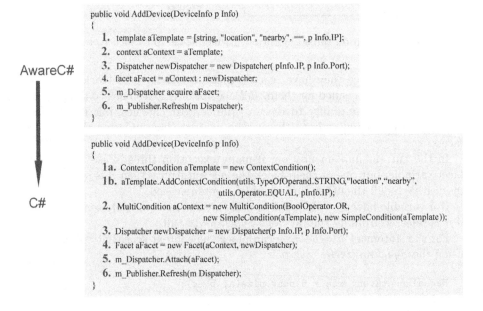

Fig. 3. Example of translation from AwareC# to C#

5.1   Support Infrastructure

The architecture needed to deploy AwareC# application consists in an infrastructure whose goal is to gather and manage the context information. The infrastructure collects information from various sensors and stores and updates it. Therefore, the infrastructure is responsible for tracking the context and for notifying the multifacet programming constructs whenever the context changes as it is illustrated in figure 2. Once a multifacet entity is notified, it checks whether its current facet must be changed.

In order to get AwareC# programs compiled we translate them into C# programs using a preprocessor which was created using ANTLR parser generator [16]. We provided the AwareC# grammatical description as an input for the

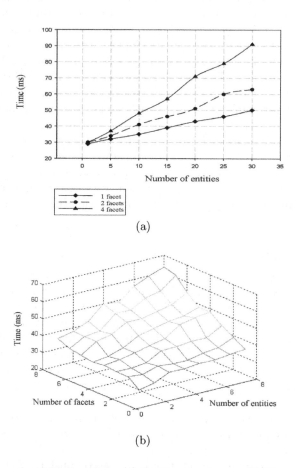

(a)

(b)

Fig. 4. Variation of notification time: (a) the time increases with the number of multifacet programming constructs and the number of facets; (b) the number of facets and the number of multifacet programming constructs almost equally influence the notification time

parser generator and get the preprocessor module as the output. This module is able to translate the code written in AwareC# into C# code (library calls). Figure 3 depicts an example of the translation from AwareC# to C#.

5.2 Performance Analysis

In the model we have proposed, an application consists both from context independent and context-aware programming constructs. The latter induces some delays, as the context-aware programming constructs need some time to react to the context changes. In this section, we evaluate the values of these delays i.e. the extra time needed by the context-aware programming constructs compared to not-context-aware constructs. In order to measure the delays we used a HP iPAQ rx3700 (400MHz) having 152MB memory. The software infrastructure is implemented in C# .NET Compact Framework. We measured the time needed for the whole collection of the context-aware programming constructs to be notified by a context change. We kept the number of facets constant while varying the number of multifacet constructs of collections as 1, 5, 10, 15, 20, 25 and 30 multifacet constructs. The results are presented in figure 4(a). While for an entity having 1 facet the notification time is around 30 ms, for a collection of 30 multifacet programming constructs each having 4 facets the notification time goes up to 90 ms. This delay is reasonable for a broad range of applications.

In figure 4(b) we compare two situations: (1) how the notification time increases when the number of facets are constant while varying the number of multifacet constructs; and (2) how the notification time increases when the number of multifacet constructs are constant while varying the number of facets. The two numbers almost equally influence the notification time as it is showed in figure 4(b).

6 Related Work

The programming model COP (Context Oriented Programming) [10] [4] has been implemented in Python. A COP compliant program is structured in context independent code and context dependent code (*open term*). For each open term many code snippets (called stubs) may be defined, one snippet for each context. During the program execution when an open term is met the stub corresponding to the current context is executed. The open term concept and its stubs are somehow similar to the multifacet programming language construct and its facets.

In our approach, we group each multifacet programming construct with its associated facets. In contrast with our approach, in COP model there is no direct link between an open term and the stubs that can fill the open term in. The lack of the direct link increases the effort needed to accomplish the matching between the open terms and the stubs. The matching process between the open terms and the stubs is based on the goals and context of that open term. Once the right stub for an open term is found out, the code of the stub is inserted into the program. This comes with an issue concerning the scoping system. The scoping mechanism proposed is presented in [10] where bound and half-bound parameters are dynamically scoped and unbound parameters exist only within the lexical scope of the stub. In our approach, the context-awareness was implemented at the constant, variable, function and object levels. In COP the granularity of the context-awareness is somehow finer as it has been implemented at statement, term, operator, method, expression, object and cast levels. It remains to be seen whether such a fine level of granularity it is really needed in the practice of the context-aware applications.

ContextL, which is an extension of CLOS, contains a context sensitive inheritance mechanism [2]. Multiple definitions are allowed for a class, one for each context (layer). In contrast with this approach we implemented context-awareness not at the class level, but at the object level. Thus we allow the dynamic modification of the context-awareness at runtime while in ContextL this cannot be achieved.

7   Conclusion

In this paper we presented our proposal regarding a context-aware extension of the C# programming language. The language extension is called AwareC# and contains a model of the context and it also contains constants, variables, functions and objects whose values and behaviors are context dependant. Basically, from programmer's point of view, the context-awareness of the multifacet constant, variable, function and objects is defined once (usually at the beginning of the program). Later, during the execution, multifacet programming constructs automatically follow the context changes by exposing and hiding their facets. When the times comes the multifacet programming constructs are ready to be used having the right facet (value / behavior) exposed. Having the context-awareness encapsulated at the programming language level decreases the programmer's responsibility in managing the context-awareness without reducing the programmer's ability to express the context-dependent behavior.

Our future work includes an investigation on the need to introduce, at the language level, a more complex model of context.

References

1. Chen, G., Kotz, D.: A Survey of Context-Aware Mobile Computing Research. Technical Report TR2000-381, Computer Science Dartmouth College (2000)
2. Costanza, P., Hirschfeld, R.: Language Constructs for Context-oriented Programming - An Overview of ContextL. Dynamic Languages Symposium (at OOPSLA'05) (2005)
3. Gu, T., Pung, H. K., Zhang D. Q.: A Service-oriented middleware for building context-aware services. Journal of network and computer applications **28(1)** (2005) 1–18.
4. Keays, R., Rakotonirainy, A.: Context-oriented Programming. Proceedings of the International Workshop on Data Engineering for Wireless and Mobile Access (2003)

5. Kniesel, G.: Dynamic Object-Based Inheritance with Subtyping. PhD thesis, University of Bonn, Institute for Computer Science III (2000)
6. Korpip, P., Mntyjrvi, J., Kela, J., Kernen, H., Malm,E.-J.: Managing Context Information in Mobile Devices. IEEE Pervasive Computing Magazine special issue: Dealing with Uncertainty **2(3)** IEEE Computer Society (2003) 42–51
7. Lieberman, H.: Using Prototypical Objects to Implement Shared Behavior in Object Oriented Systems. In Proceedings of OOPSLA6 (1986) 214–223
8. Mostfaoui, G. K., Pasquier-Rocha, J., Brezillon, P.: Context-Aware Computing: A Guide for the Pervasive Computing Community. In Proceedings of the 2004 IEEE International Conference on Pervasive Services (ICPS004) (2004) 39–48
9. Oh, Y., Woo, W.: A Unified Application Service Model for ubiHome by Exploiting Intelligent Context-Awareness. In Proceedings of the 2nd International Symposium on Ubiquitous Computing Systems (UCS 04), Tokyo, Japan (2004) 117–122
10. Keays R.: Context-Oriented Programming. University of Queensland (2002)
11. Roman, M., Hess, C., Cerqueira, R., Ranganathan, A., Campbell, R.H., Nahrstedt, K.: A Middleware Infrastructure for Active Spaces. IEEE Pervasive Computing **1(4)** (2002) 74–83
12. Schilit, B.N., Theimer, M.: Disseminating active map information to mobile hosts. IEEE Network **8(5)** (1994) 22–32
13. Shehzad, A., Ngo, H. Q., Pham, K. A., Lee, S.Y.: Formal Modeling in Context-Aware Systems. In Proceedings of KI2004 Workshop on Modeling and Retrieval of Context, University of Ulm (2004)
14. Weiser, M.: The Computer for the 21st Century. Scientific American **265(3)** (1991) 94–104
15. Yau, S. S., Karim, F.: An Adaptive Middleware for Context-Sensitive Communications for Real-Time Applications in Ubiquitous Computing Environment. Real-Time Systems **26(1)** (2004) 29–61
16. http://www.antlr.org/ (available May 2006)

Toward Wide Area Interaction with Ubiquitous Computing Environments

Michael Blackstock[1], Rodger Lea[2], and Charles Krasic[1]

[1] Department of Computer Science, University of British Columbia
201-2366 Main Mall, Vancouver, B.C., Canada
[2] Media and Graphics Interdisciplinary Centre, University of British Columbia
FSC 3640 - 2424 Main Mall, Vancouver, B.C., Canada
{michael@cs, rodgerl@ece, krasic@cs}.ubc.ca

Abstract. Despite many years of ubiquitous computing (ubicomp) middleware research, deployment of such systems has not been widespread. We suggest this is in part because we lack a shared model for ubicomp environments, and that most existing systems are constrained to single administrative (and network) domains. To address this, this paper presents work in progress toward a core common ubicomp environment model derived from an analysis of several existing ubicomp systems. This model lends itself to interoperability across domains, and for use in a middleware platform used to adapt existing ubicomp systems to this common model for wide area access. This platform design, based on enterprise application integration (EAI) approaches, highlights the benefits of Web Services and Semantic Web technologies for exposing ubicomp environments to applications outside the administrative domain.

1 Introduction

Over the last several years, researchers have created middleware, toolkits and operating systems to support the development of integrated ubicomp environments in homes, meeting rooms and other environments. While these contributions have matured to address many challenges in supporting ubicomp application development, they have mostly been confined to lab prototypes. Many reasons have been cited for this lack of progress [1]. One reason may be that it is difficult to persuade others to use a non-standard technology; perhaps there is a perception that software quality and ongoing support from their creators as research prototypes will be questionable [2]. Even when researchers have attempted to leverage open standards such as Web protocols, they still have not been widely adopted.

We argue that there are two important and related impediments to wider deployment of ubicomp environments. One is that, until recently, system designers have focused primarily on supporting applications and user access within single administrative or network domains. While many researchers have suggested the use of a common systems infrastructure [3-5] to address this, none so far have been adopted as a standard. Because ubicomp systems have evolved from a range of established middleware platforms, and each has its own strengths for application development, heterogeneity at the interface, communications, and abstraction levels should be

P. Havinga et al. (Eds.): EUROSSC 2006, LNCS 4272, pp. 113–127, 2006.
© Springer-Verlag Berlin Heidelberg 2006

expected. It is likely too early to agree on a single approach to building ubicomp applications. That said, until systems can support integration with devices, services and sensor information across domains they will not be widely adopted in the "real world". Meaningful cross-domain integration also relies on a shared understanding of the interactions that may occur between applications and the supporting infrastructure. A second impediment therefore, is the lack of a shared model for ubiquitous computing environments. We need a solution that allows heterogeneity with a single environment while supporting a degree of interoperability across domains.

Our approach is to create a bridge between existing systems to a common model suitable for wide area interaction, with the goal that this model may lead to an interoperable standard. To build this bridge, we can look toward progress in other domains such as enterprise application software development and integration (EAI). Conventional enterprise middleware technology has evolved rapidly from single tier mainframe applications to the use of shared objects and *message brokers* to integrate heterogeneous systems across an organization. Until recently, a lack of standards at the middleware and component levels between domains has hindered inter-enterprise communications and interoperability. To address this, the enterprise software development community has turned toward the use of Web Services and associated standards efforts. When we compare the evolution of ubiquitous computing environments with that of EAI, they seem to be following a similar trend. Just as *internal middleware* for enterprise application development matured for single domains followed by the use of *external middleware* for integration, we propose that the wider deployment of ubicomp systems for access by applications in other domains will require similar external middleware for interoperability.

Toward addressing cross-domain ubicomp interoperability, this paper describes work in progress for a shared model for ubiquitous computing environments and the design of a flexible external middleware platform based on Web Services to adapt existing ubicomp systems to this model. This shared model must lend itself to interoperability and specialization for environment domains such as the home, the office and public places and adaptation to existing ubicomp system abstractions.

The remainder of this paper is organized as follows. In section 2 we present an analysis of some representative ubicomp systems toward deriving a common set of abstractions. Our core model is then described in Section 3. Section 4 presents our design and initial prototype of our external-middleware system called Ubicomp Integration Framework. Section 5 discusses related work, and section 6 concludes with a discussion of future work.

2 Model Analysis

To drive our analysis, we envision scenarios where mobile phone applications connected to wide area networks interact with public ubicomp environments such as shopping malls and museums. We also anticipate that application servers hosted outside of a ubicomp environment's network domain need to make use of resources there. This can occur across a large university campus or between organizations to link smart meeting rooms for example. To support these scenarios, applications hosted outside of a ubicomp environment's home network domain must interact with resources in another environment across an administrative and/or network domain boundary.

Our goal is to derive a model that finds the right balance between interoperability and suitability for cross domain interactions while maintaining some of the flexibility of a given underlying ubicomp system. A new level of abstraction however, can make application development more difficult, especially if it doesn't match the problem at hand. Just as different programming languages and supporting libraries support some application domains better than others, we expect that different environment models will need to co-exist. Since we do not expect all local ubicomp applications to require cross-domain interaction, we need not replace an existing set of abstractions and associated APIs for native application development; rather we can provide an interoperable model as an alternative more suitable for wide area application access.

To derive our model for wide area interoperability we have begun with an analysis of existing ubicomp systems [3-15] to find some common ground. Through this analysis we've arrived at the following taxonomy of abstractions suitable for capturing many of the semantics exposed by many existing systems.

- **Environment Model** that encapsulates the current state of the environment including the components available, the types of context and services components provide and other static and dynamic aspects of the environment. This is often handled by a service discovery system, but in some systems, a component that handles more complex models of the environment is provided.
- **Entities** are base-level abstractions such as people, places, things, groups and activities that can be extended to environment-specific entities such as game players, living rooms, PDAs and meetings.
- **Context** associated with entities. Context information can include values such as location, temperature, direction, or time, or higher level inferred context such as user activities and goals.
- **Services** or functionality associated with entities.
- **Entity Relationships** such as geometric, social, and activity-related relationships between people, places and things.
- **Events** that entities can fire related to a change of state of an entity, for example a door closing, a light turning on or a slide change in a presentation. Events can also be fired by services, context, or due to entity relationship changes. Several platforms support events as separate abstractions in their own right.
- **Data or Content** related to an entity. For example, this could include a user's personal notes, or documents a group has been working on associated with a meeting.

At first glance, the notion of an *application* itself may be one of the abstractions that need to be considered since they can be embedded in the ubicomp environment. Such is the case with Aura for example [3] where an application or user *task* is represented as a collection of abstract services that can be transferred from environment to environment. Since our aim is to provide an application independent model of an environment we do not include the notion of an application explicitly in our model.

Table 1 summarizes our analysis indicating where a ubicomp system has explicitly exposed one of the abstractions listed.

Table 1. Abstractions exposed by existing ubiquitous computing and context aware systems. ✓ = abstraction exposed, X = not exposed, P = partially or implicitly exposed by the system.

Systems	Env. Model	Entities	Context	Inferred Context	Services	Entity-Entity Rel.	Events	Content/ Data Storage
Cooltown	X	✓	P	X	P	✓	P	P
Context Toolkit	P	✓	✓	✓	✓	X	P	X
JCAF	X	✓	✓	✓	P	✓	✓	X
iROS	P	X	✓	X	✓	X	✓	✓
Gaia	✓	X	✓	✓	P	X	✓	✓
Active Campus	✓	✓	✓	P	✓	✓	P	P
Aura	P	P	✓	X	✓	P	P	✓
EasyLiving	✓	✓	✓	P	P	✓	P	X
CoBrA	P	P	✓	✓	P	P	P	X
Sentient Computing	✓	✓	✓	P	P	✓	P	X
Sentient Objects	P	P	✓	✓	P	P	✓	X

We have also noted that there are three related aspects to an environment model that must be considered. The *Environment State* aspect consists of entities, entity relationships, and the current state of those entities: current context values and content for example. Secondly, the *Environment Meta-State* consists of entity instances associated with the *types* and *quality* of events, services, context and content. This aspect is required to support introspection. Both the Environment State and Meta-State must be exposed to applications to query the capabilities of an environment and make use of them. Finally the *Environment Implementation* links entity instances to the specific components that supply the services, context and events associated with entities. These aspects depend on one another and may change over time. The Meta-State depends on the current Implementation, and the current State depends on the Meta-State. In some cases, components associated with an entity such as a mobile user can depend on the current situation; the Implementation can depend on the Environment State. We will further elaborate on these three aspects in Section 3.

To conclude our analysis, we have derived the following requirements for a common environment model suitable for wide area application interaction.

- The model should support **interoperability** between different environment types such as the home, the office and public places while also supporting **specialization** for these different environments. This will allow environment-specific capabilities while allowing very general purpose applications to work across locations.
- The model should lend itself to a relatively **straightforward mapping to existing ubicomp systems' abstractions**. This means finding the right tradeoff between being suitably generic across a wide range of systems but semantically close enough to specific systems for flexibility.
- A common model must support **introspection**, exposing not only the current environment state, but also its current capabilities such as service interfaces and the types of context associated with entities. This will allow applications to query the environment for its current capabilities.

- To support wide area interactions, the model should lend itself to a simple and **coarse-grained access interface** for applications to minimize the communications required over wide area networks. One method call should accomplish as much as possible.
- The model should **separate exposed abstractions from implementation abstractions**. This separation of concerns will allow implementation independence, and dynamic binding of components to entities without application involvement. Implementations can change depending on the current context, and systems can support component failover for greater reliability.

In addition to these model requirements we claim that this model should be *executable*, that is, have the ability to be queried and reasoned with [16, 17]. Support for flexible queries will allow applications to discover entities and associated services, and allow applications to determine whether their requirements can be satisfied. With integrated reasoning a supporting system can maintain the exposed model as its composition changes, simplify integration tasks, and provide missing general purpose capabilities such as context interpretation and entity aggregation for example.

3   Core Model

In this section we describe the three aspects of our core model illustrating the relationships between entities and other abstractions in an Environment State, Meta-State and Implementation followed by a simple example to show the relationship between these aspects.

3.1   Environment State and Meta-state

The Environment State consists of entity instances, their relationships, and their current context and content values. These abstractions and their relationships are shown in Figure 1 (a). The *Environment* object serves as the root entity of an environment and *hosts* other *Entities* and subclasses of Entities such as places, people and devices. Context values are retrieved by requesting the value of a *contextAttribute*. The contextAttribute property and *ContextValue* object may be specialized as shown to support different data types. Entities may also have content associated with them as shown by the *hasContent* relationship with a *Content* object.

The Environment Meta-State aspect associates Entities with the types and quality of events, services, context and content to support introspection as shown in Figure 1 (b). When an Entity has context associated with it, the *hasContext* property will refer to a *ContextType* object. This object will refer to a *contextAttribute* property used when retrieving a *ContextValue*. The ContextType may include properties to specify the quality of this context as shown. Similarly, entities may *expose* certain *ServiceInterfaces*. Clients of the model can then call these services as specified in a *ServiceDescription* object. ServiceDescriptions can be specialized to support standard service descriptions such as Web Services Description Language WSDL [18]. The type of events that may be fired by an Entity may also be specified using the *fires* property and an *EventType* object.

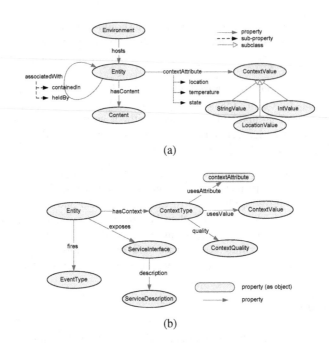

(a)

(b)

Fig. 1. Environment State (a) and Meta-State (b) aspect abstractions and relationships

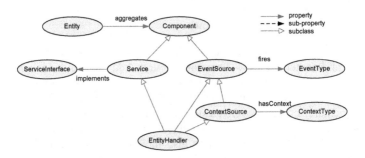

Fig. 2. Environment Implementation model aspect

3.2 Environment Implementation

To avoid dealing with a plethora of sensors, actuators services, and software components, we've found that ubicomp systems typically expose the following high level component types.

- **EventSources.** Software components responsible for firing events related to one or more entities. An event may be related to a location change, or moving to the next slide of a presentation for example. These components correspond to the Event abstraction exposed by the iROS EventHeap or Gaia Event Manager.

- **ContextSources.** Components responsible for delivering context, either inferred high level context, or low level sensor data related to one or more entities. These components can be queried or can fire events; they are also EventSources. These components correspond to sensor input components such as Context Widgets in the Context Toolkit.
- **Services.** Components responsible for delivering functionality associated with one or more entity abstractions and their exposed ServiceInterfaces. These components typically expose a set of method calls. For example, storage and retrieval such as the iROS DataHeap and the Gaia Context File System are considered a specialized service component.
- **EntityHandlers.** Components responsible for the complete manifestation of one or more entities. These components deliver all of the associated services, context, content and events related to these entities. An entity handler may be used to adapt a Context Toolkit Aggregator or the *web presence* software representing people, places or things in Cooltown for example.

These component abstractions are captured in our Implementation aspect of an Environment as shown in Figure 2. Here we show that components are *aggregated* by an Entity instance and form a class hierarchy as shown. *Service* components *implement* a ServiceInterface, *EventSource* components *fire* EventTypes, and *ContextSources* have a ContextType. When an entity aggregates a component, it exposes the types and interfaces that component supports in the Environment Meta-State. These component abstractions can be used to map the common model to corresponding APIs in an existing system to invoke services, retrieve context or fire events.

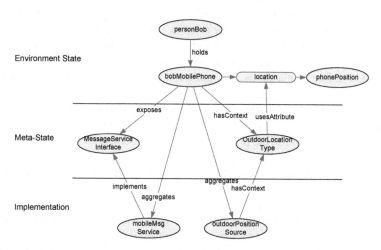

Fig. 3. Model example illustrating dependencies between Environment State, Meta-State and Implementation aspects

3.3 Model Example

The three aspects of the Environment model and how they relate to one other are illustrated by a simple example shown in Figure 3. A person Entity *personBob* is shown to hold a mobile phone *bobsMobilePhone*. In the current Environment State, this phone has a *location* called *phonePosition* containing the longitude and latitude of the device. In the current Meta-State, bobsMobilePhone is shown to *expose* a *MessageServiceInterface* and has the *OutdoorLocationType* of context available for applications. The OutdoorLocationType uses the *location* attribute as shown.

In the current Implementation, bobsMobilePhone enitity aggregates the *mobileMsgService* Service component and the *outdoorPositionSource* ContextSource component. The mobileMsgService *implements* the *MessageServiceInterface* type, and the outdoorPositionSource *hasContext OutdoorLocationType* as shown. This implies that the bobsMobilePhone exposes the same ServiceInterface and ContextType to applications.

4 Model Discussion

With a shared core model as described, some degree of interoperability is possible. Applications can browse an environment by entity relationships, and display the types of context and services associated with these entities for example. A higher degree of interoperability is only possible, however, when applications share a deeper semantic understanding of entities and their associated resources with the supporting infrastructure. To address this we propose the use of *Environment Profiles* to specialize the core model for specific environment types. These profiles can specify the classes of entities, services, context, events, content and their possible relationships for a given environment type. A home profile, for example, can consist of typical places in the home such as kitchens and living rooms, device types such as appliances, and home entertainment systems. Home context and services can include temperature, lighting controls and room-resolution location sensors. An application interacting with the home environment can then "turn the lights on in a room" when a user arrives by specifying the expected lighting service associated with a room. Similarly a museum profile could define displays, visitors, galleries, display content, visitor location, and interests.

With a supporting system, applications can query an environment for the profiles it supports. Profiles may be extended further by an integrator to provide extensions specific to a deployment, at the possible expense of interoperability. Through the use of a specialized core model, and supporting infrastructure to map this model to existing systems we argue that it is possible for applications hosted outside of an environment's local domain to interoperate with an environment's resources independent of the ubicomp middleware used.

Based on our analysis, our model will lend itself to a straightforward mapping to a range of existing systems, however we do anticipate that some flexibility will be lost in providing this new level of abstraction that we intend to qualify in future work. At best we can mirror the "native" environment model presented by existing systems to external applications but in some cases we may need to present an alternative wide area model. The Meta-State aspect of the model supports introspection for applications to determine whether its requirements can be met by the environment. Through environment and entity aggregation, the model lends itself to the use of a coarse grained interface, for example, to access the current state of several entities using a single query rather than accessing components directly. The separation of exposed State, Meta-State and Implementation aspects in the model itself allows a supporting system to vary component aggregations independently of the exposed abstract context types or service interfaces. Finally, by expressing this conceptual model explicitly using semantic web technologies such as the Resource Definition Framework (RDF) [19] we can support model queries and reasoning using a supporting system.

5 Integration Framework

An open question is whether our core model adequately captures the semantics of a range of underlying system abstractions for wide area application development. To address this we have begun implementation of a middleware platform called Ubicomp Integration Framework (UIF) designed specifically to adapt existing ubicomp environment middleware to a common model.

Our requirements for a suitable integration framework are that it is general purpose enough to adapt a wide variety of existing systems to our common model. The system should provide a standard interface suitable for wide area environment introspection and interaction and support dynamic reconfiguration of the exposed model when the underlying ubicomp system(s) change. Our approach is to leverage Web Services for application-environment interaction. We have also used semantic web tools and in particular the use of RDF and the Web Ontology Language (OWL) [20] to describe our model ontologies and instance data. A key component of the current system is an integrated reasoning engine to maintain all three aspects of our environment model and support flexible queries about the environment composition.

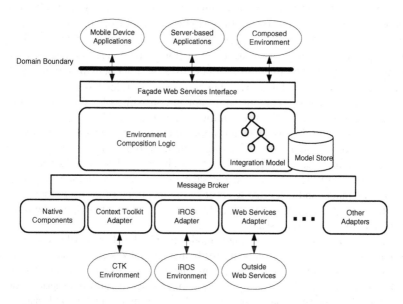

Fig. 4. Proposed high-level ubicomp integration framework architecture

5.1   Framework Architecture

A high level view of the UIF architecture is shown in Figure 4. The Façade Web Services interface allows applications outside of the environment domain to interact with the environment's resources as described in our model. It calls the Environment Composition Logic (ECL) which provides a comprehensive API to query, and interact with the environment model. The Environment Composition Logic delegates model-related calls directly to the Integration Model subsystem to manage the environment model on its behalf. The Integration Model and its store maintain the relationships between entities, their context, events, exposed services, and the components using our core model. The Integration Model may be queried to find entities by type, entity relationships, the services, or context they currently expose for example. It is also used by the ECL to determine the destination Adapter for messages to access context, invoke services or subscribe to events using the appropriate internal middleware platform.

Adapters map internal ubicomp middleware implementations to the common model effectively hiding the heterogeneity of the internal ubicomp system used. For example, a Context Widget in the Context Toolkit may be mapped to a ContextSource component and a Service component in our shared model. An iROS event may map to an EventSource in the model. To support loose coupling between Adapters and the rest of the framework the ECL communicates with adapters using a standard message broker. Smaller systems may not require a message broker, and a simpler alternative such as an RPC mechanism like RMI or a centralized Java framework may be used between Adapters and the rest of the system.

5.2   Prototype Implementation

In our initial prototype we've implemented the Web Services Façade, Environment Composition Logic, and the Integration Model in a proof of concept. A simple test application accesses the model exposed by the UIF using web services. This prototype is implemented using a Tomcat server [19] with the Axis reference Web Services implementation [20]. The Integration Model uses the Jena Semantic Web engine [21] to parse, store OWL and RDF knowledge bases, and for rules-based reasoning.

The prototype models a campus environment consisting of a single static place entity (campusPlace) and mobile user entities located there. User entities are added to the model when a user logs in to the application. Each user entity exposes a ServiceInterface to send and receive messages, their current indoor or outdoor position, and an event that is fired when their location changes. The campusPlace entity exposes composite services to send a message to "geocast" a message all users contained in this place, and a service to retrieve the locations of all users there. The test application presents a browser user interface to mobile users to display the position of users on a map, and send messages.

On system startup, the environment model is initialized from files containing our core model ontology, an environment profile that specializes core objects, and instance data described using OWL. Files containing integration rules and a standard set of rules for OWL inference are then read by the system. One set of rules is generic to our integration model, while the other is specific to the environment profile and instance data. These rules ensure that entity-entity and component-entity relationships are maintained as new entities and components are added or removed. For example, the following rule states that if an entity e aggregates a service component s, and that service component implements an interface si, then the entity itself exposes this interface.

$$\forall e, s, si : \exp oses(e, si) \leftarrow Service(s) \land aggregates(e, s) \land implements(s, si)$$

Similar rules are used to expose context and event types for aggregated ContextSources and EventSources. Such entity-component aggregations can be specified statically by an integrator, by an adapter when a new component is added, or inferred by other rules. Other integration rules can take into consideration the attributes of entities when aggregating components such as their types or locations they serve. For example the following rule will aggregate a context source for providing indoor position information (*csIndoorPositionSource*) only when a *Person* entity e is *contained in* the *csBuilding* place.

$$\forall e : aggregates(e, "csIndoorPositionSource") \leftarrow Person(e) \land containedIn(e, "csBuilding")$$

As a result of this rule, a component aggregation in the Implementation aspect of the model depends on the Environment State, that is, the current position of a Person.

The reasoning engine also supports introspection of the environment to find entities that match a certain criteria such as the service interfaces exposed or context types supported. The environment model can be queried using the Simple Protocol and RDF Query Language (SPARQL) [21]. For example, to find the entities that expose a *campus:MessageService* service interface, applications can issue the following query:

```
SELECT ?entity WHERE
{?entity uif:exposes campus:MessageService}
```

Since the Person entity *campus:bob* and the Place *campus:campusPlace* both expose this interface, the following is returned:

```
( ?entity = campus:bob )
( ?entity = campus:campusPlace )
```

In an effort to assess the work involved in integrating existing ubicomp systems, we've also begun work on adapters for the Context Toolkit (CTK) [12], iROS [6] and the Equator Component Toolkit (ECT) [22] for our system. The iROS adapter was abandoned early since the available open source version is no longer maintained while work on adapters for CTK and ECT continues.

5.3 Discussion

From our initial prototype work we have found that semantic web technology is useful for expressing our environment model. Integrators can specify static entities and components that are linked using sets of integration rules. The system can then add and remove entities and components that are then automatically incorporated into the three aspects of our environment model. The integrated reasoning engine can also help maintain the environment model as it changes, maintain consistency, and perform validation checks. The same engine may be used to change component aggregations based the current situation, and as a general purpose context interpreter [9, 10].

Using SPARQL, clients of the environment can query for entities by type, that expose specific services, by entity relationship or other attributes. Semantic web standards also allows integrators to take advantage of RDF [19] and OWL tools available for creating and maintaining environment profiles and descriptions. We note that recently the use of semantic web technologies has been applied toward software engineering and has been recognized as a potentially valuable tool to facilitate the management of components in application servers and Web Services middleware [16].

Wrapping an environment using web services has several advantages. Applications and user interfaces can be developed independently of the environment, allowing environment integrators to focus on the resources their environment exposes without concern for user interface issues. It allows application logic to be hosted on either on a mobile device, or a server as necessary. A centralized server also provides a natural opportunity to address issues such as security, privacy and trust, which are especially important for environments exposed to applications outside the administrative domain. We hope to explore these issues in future work.

Like the task of integrating various sensors with any ubicomp middleware platform, integrating a given ubicomp environment under a common model can be challenging. Our ongoing integration work has informed our adapter interface and clarified the typical mapping abstractions needed between existing systems and our model. We expect that adapter development may be streamlined through the use of our framework and associated tools that we expect to address in future work.

6 Related Work

Several ubicomp systems focus on presenting a unified model of ubicomp environments for application development. Shilit et al. described a model where context information was maintained by a set of variables handled by environment servers representing entities such as people, places or groups [23]. The Sentient Computing project infrastructure [14] maintained an integrated model of the environment in a relational database for application development. Ontology-based systems model context as relationships between entities and context objects using semantic web technologies [9, 10, 24] as we do. Henricksen et al. present a formal model of context information independent of an implementation that captures the temporal aspects of context and context quality [25]. While others have attempted to provide abstractions for application development, through our survey and analysis, we have proposed to unify the abstractions exposed by existing ubicomp systems to provide a *coarse grained* interface for applications interacting with environments across domains. The core model described brings together notions of entities, entity relationships, context, services, and events available in an environment. It then links these exposed abstractions with implementations components to facilitate mapping these exposed abstractions to an underlying ubicomp system.

Like ActiveCampus [11], our UIF design should be suitable for access to large scale environment services using web browsers on mobile devices with an appropriate proxy server. In contrast to ActiveCampus, our UIF design is not intended only for large scale environments, but also environments such as museums and offices. The

UIF design exposes environment resources independent of the user interface and supports applications running on both servers and mobile devices using web services.

Recently the use of Grid technologies has been proposed to support wide area ubiquitous computing [2, 26]. While we agree that this approach shows promise, the level of coordination between wide area applications and ubiquitous computing environments may not always require a Grid infrastructure. As the need for integration between many ubicomp systems increases, the Grid as an extension of Web services and the Semantic Web will likely play a key role in environment integration.

The ReMMoC [27] system is a device-side *reflective* or *adaptive* middleware system that modifies its behaviour by means of inspection and adaptation. Like the UIF, this system uses Web Services to isolate applications from the heterogeneity of services that may be available. Friday et al. [28] describe requirements for an infrastructure that hides the heterogeneity of various service platforms such as UPnP [29] and HAVi [30] to address a number of open issues related to service deployment for ubiquitous computing. The UIF differs from ReMMoC in that it is infrastructure-based, and unlike both of these systems, our proposed model and supporting system strives to provide a set of abstractions for an integrated ubicomp environment suitable for cross-domain interoperability built on Web Services.

7 Conclusions and Future Work

In this paper we have proposed solutions to two related impediments to the wider deployment of ubicomp systems. The first is a core common model for ubicomp environments based on abstractions exposed by existing ubicomp systems. This entity-centric model separates the state and meta-state aspects of an environment model from the implementation concerns. The second is the design of the Ubicomp Integration Framework (UIF), *external middleware* designed to adapt existing ubicomp systems to this common model for wide area access. An early prototype has highlighted several advantages of the UIF design including the use of semantic web technologies in the specification and maintenance of the exposed model. Creating adapters for existing systems looks challenging but we are encouraged by the possibility of integrating several ubicomp systems to enable some degree of interoperability between them.

Since our derived abstractions are based on an analysis of several existing systems, we expect that a model based on them will service a wide range of applications. However, it is not yet clear that our model is ideal for *wide area* applications. By building more comprehensive prototypes we expect to understand better the considerations for wide area interactions between applications and supporting infrastructure.

The introduction of a new set of abstractions will (necessarily) hide the details of the underlying ubicomp environment, but likely at the cost of flexibility. Where there is a close match between our proposed abstractions and the underlying system, we can expect to mirror many of its capabilities to outside applications, but in some cases, there will not be an obvious match between one or more abstractions in our model and the "native" model of a given system. In these cases our proposed model may not lend itself to a particular approach to user mobility, application development, or may not provide the level of performance or control needed for a specific application. For example, it may be difficult to integrate the Aura's notion of a "user task" in our model, impractical to intercept mouse movement events from the iROS EventHeap, or deliver high bandwidth multimedia content. In these cases we may need to be satisfied with presenting an alternative wide area model that co-exists or provides hooks into the native environment model for local applications.

From our initial experience we have come to realize that maintaining a dynamic environment model will also be a challenge as the composition of entities, components and the context of entities in an environment changes. To address this we will expand the use of the integrated knowledge base and associated reasoning engine to evaluate its suitability for dynamically maintaining the three aspects of our model. Finally we intend to qualify our UIF system's capabilities as a stand-alone middleware platform for large scale public ubicomp deployments where wide area interaction between applications and supporting environments is critical.

References

1. Davies, N., Gellersen, H.-W.: Beyond Prototypes: Challenges in Deploying Ubiquitous Systems. IEEE Pervasive Computing, Vol. 1 (2002) 26-35
2. Storz, O., Friday, A., Davies, N.: Towards 'Ubiquitous' Ubiquitous Computing: an alliance with the Grid. System Support for Ubiquitous Computing Workshop at the Fifth Annual Conference on Ubiquitous Computing (UbiComp 2003), Seattle (2003)
3. Sousa, J.P., Garlan, D.: Aura: an Architectural Framework for User Mobility in Ubiquitous Computing Environments. Proceedings of the 3rd IEEE/IFIP Conference on Software Architecture. Kluwer, B.V. (2002)
4. Roman, M., Hess, C., Cerqueira, R., Ranganathan, A., Campbell, R.H., Nahrstedt, K.: A Middleware Infrastructure for Active Spaces IEEE Pervasive Computing 1 (2002) 74-83
5. Kindberg, T., Barton, J., Morgan, J., Becker, G., Caswell, D., Debaty, P., Gopal, G., Frid, M., Krishnan, V., Morris, H., Schettino, J., Serra, B.: People, places things: Web presence for the real world. Third IEEE Workshop on Mobile Computing Systems and Applications Monterey, California (2000)
6. Ponnekantia, S.R., Johanson, B., Kiciman, E., Fox, A.: Portability, extensibility and robustness in iROS. Proceedings of IEEE International Conference on Pervasive Computing and Communications, Dallas-Fort Wirth (2003)
7. Salber, D., Dey, A.K., Abowd, G.D.: The context toolkit: aiding the development of context-enabled applications. Proceedings of the SIGCHI conference on Human factors in computing systems. ACM Press, Pittsburgh, Pennsylvania (1999)
8. Brumitt, B., Meyers, B., Krumm, J., Kern, A., Shafer, S.A.: EasyLiving: Technologies for Intelligent Environments. Proceedings of the 2nd international symposium on Handheld and Ubiquitous Computing. Springer-Verlag, Bristol, UK (2000)
9. Chen, H., Finin, T., Joshi, A., Kagal, L., Perich, F., Chakraborty, D.: Intelligent Agents Meet the Semantic Web in Smart Spaces. IEEE Internet Computing 8 (2004) 69-79
10. Gu, T., Pung, H.K., Zhang, D.Q.: Toward an OSGi-Based Infrastructure for Context-Aware Applications. IEEE Pervasive Computing 3 (2004) 66-74
11. Griswold, W., G., Boyer, R., Brown, S., W., Truong, T.M.: A component architecture for an extensible, highly integrated context-aware computing infrastructure. Proceedings of the 25th International Conference on Software Engineering. IEEE Computer Society, Portland, Oregon (2003)

12. Dey, A.K.: Providing Architectural Support for Building Context-Aware Applications. PhD Thesis. College of Computing, Georgia Institute of Technology (2000)
13. Bardram, J.E.: The Java Context Awareness Framework (JCAF) - A Service Infrastructure and Programming Framework for Context-Aware Applications. Pervasive Computing: Third International Conference. Springer Berlin / Heidelberg, Munich, Germany (2005)
14. Addlesee, M., Curwen, R., Hodges, S., Newman, J., Steggles, P., Ward, A., Hopper, A.: Implementing a sentient computing system. IEEE Computer **34** (2001) 50-56
15. Biegel, G., Cahill, V.: A Framework for Developing Mobile, Context-aware Applications. Second IEEE International Conference on Pervasive Computing and Communications (2004) 361
16. Tetlow, P., Pan, J.Z., Oberle, D., Wallace, E., Uschold, M., Kendall, E.: Ontology Driven Architectures and Potential Uses of the Semantic Web in Systems and Software Engineering. W3C (2006) http://www.w3.org/2001/sw/BestPractices/SE/ODA/060211/
17. Oberle, D., Eberhart, A., Staab, S., Volz, R., Jacobsen, I.H.-A.: Developing and Managing Software Components in an Ontology-based Application Server. Middleware 2004, ACM/IFIP/USENIX 5th International Middleware Conference, Vol. 3231. Springer, Toronto, Ontario, Canada (2004) 459-478
18. Christensen, E., Curbera, F., Meredith, G., Weerawarana, S.: Web Services Description Language. W3C (2001) http://www.w3.org/TR/wsdl
19. Resource Description Framework. http://www.w3.org/RDF/
20. Web Ontology Language (OWL) Overview. http://www.w3.org/TR/owl-features/
21. W3C: SPARQL Query Language for RDF. W3C (2005) http://www.w3.org/TR/rdf-sparql-query/
22. Greenhalgh, C., Izadi, S., Mathrick, J., Humble, J., Taylor, I.: ECT: a toolkit to support rapid construction of ubicomp environments. Proceedings of UbiComp '04 (Demonstration). Springer, Nottingham (2004)
23. Shilit, B.N., Theimer, M.M., Welch, B.B.: Customizing Mobile Applications. USENIX Symposium on Mobile and Location-Independent Computing (1993)
24. Tan, J.G., Zhang, D., Wang, X., Cheng, H.S.: Enhancing Semantic Spaces with Event-Driven Context Interpretation (2005)
25. Henricksen, K., Indulska, J., Rakotonirainy, A.: Modeling Context Information in Pervasive Computing Systems. Proceedings of the First International Conference on Pervasive Computing. Springer-Verlag (2002)
26. De Roure, D., Hey, T., Trefethen, A.E.: Where the Grid meets the Physical World - Research Issues in Grid and Pervasive Computing. (2005) http://www.semanticgrid.org/documents/gridperv3.pdf
27. Grace, P., Blair, G.S., Samuel, S.: A reflective framework for discovery and interaction in heterogeneous mobile environments. SIGMOBILE Mob. Comput. Commun. Rev. **9** (2005) 2-14
28. Friday, A., Davies, N., Wallbank, N., Catterall, E., Pink, S.: Supporting service discovery, querying and interaction in ubiquitous computing environments. Wirel. Netw. **10** (2004) 631-641
29. Universal Plug and Play (UPnP) Standards. http://www.upnp.org/standardizeddcps/default.asp
30. HAVi: Home Audio Video Interoperability. http://www.havi.org/home.html

Maintaining a World Model in a Location-Aware Smart Space

R.K. Harle

Computer Laboratory
University of Cambridge, UK
Robert.Harle@cl.cam.ac.uk

Abstract. Location-aware smart spaces fuse information from a location system and a computational world model to make contextual inferences. To date, research has concentrated on the development of cheap, realisable, accurate location systems. Comparatively little research has addressed the issue of how to maintain the world model so crucial to contextual interferences. This paper details a framework to autonomously monitor a world model (the computer-readable representation of the world) for fine-grained location systems by regularly determining its consistency with the real world. It deals with the interpretation of information that can be derived from positioning systems and the construction of a general grid-based method for evaluating consistency between real and virtual worlds. Results to date are presented using location data from an extensive deployment of a location system, before future avenues of research are identified.

1 Introduction

Location-aware computing is becoming an increasingly popular research area with the continuing miniaturisation of sensors and development of data processing techniques. Smart spaces based on real-time location data are fast emerging, introducing new conveniences and capabilities into the environment.

Location-aware computing emerged in the 1990s with the appearance of the Active Badge [1] and the PARCTAB ubiquitous computing experiment [2]. Today the authors occupy a smart space that uses a high-accuracy, pervasive, ultrasonic location system (the *Bat system* [3]) to provide positions accurate to approximately 3cm for active tags ('Bats') that can be worn by personnel or attached to objects. Location data is interpreted with reference to a world model that attempts to represent important objects in the physical space.

In dealing with such fine-grained positional data on a daily basis, it has become apparent that the problem of maintaining the world model is non-trivial. Failing to correctly update the model has a significant impact on the use of location-aware services in the building. As an example, consider hot-desking, where a user's computer desktop follows them from machine to machine as they move through the office; their nearest machine is determined dynamically using spatial indexing algorithms and their desktop transported to it using the VNC protocol [4]. Unfortunately, users occasionally shift their furniture and workstations, inadvertently invalidating the world model. Spatial containment is then incorrect for their workstation and hot-desking fails to work as

P. Havinga et al. (Eds.): EUROSSC 2006, LNCS 4272, pp. 128–142, 2006.

intended. Naturally, users choose not to rely on the system, preferring to manually transport their desktops. The resultant reduction in benefit discourages users from wearing Bats, which has the knock-on effect of reducing the system's utility for other users (if the majority of group members do not regularly wear their Bat, useful location services such as the querying of user location are devalued since the output is neither reliable nor authoritative). This paper presents a novel approach to maintaining such world models using a combination of redundant location data and mobile sensor platforms.

Perhaps the simplest solution to monitoring objects in a location-aware space is to use the same technology as is used to track people. Unfortunately this approach suffers from a number of problems, related to the present need to tag or otherwise augment a human in order to track accurately:

- The objects may not be suited to the attachment of tags—an iPAQ handheld computer is made significantly more cumbersome by the addition of a Bat, for example.
- It is expensive in terms of both time and money to deploy the number of tags required to monitor every host, device, etc.
- It is non-trivial to maintain deployed active tags, with their need for a power source.
- Some tag systems cannot offer accurate pose determination without leveraging a multi-tag solution.
- Without customising each tag to minimise visibility, the aesthetics of the space are adversely impacted.

The logical extension to this approach is to distribute specialised monitoring sensors throughout the environment. Such a sensor network would be required to recognise objects and determine their three-dimensional pose. Without augmenting the objects themselves the most promising type of sensor would be based on vision (see [5, 6] for examples of human tracking, and [7, 8] for examples of object tracking using markers). However, even if vision-based tracking is assumed to offer sufficient accuracy and coverage, there remains an issue of privacy. Any system that can be used to track untagged objects accurately can be used to track people without their consent (the authors have traditionally avoided this contentious issue by using tag-based systems to allow users to choose when they are tracked). Worse still, the pervasive deployment of video capturing devices introduces the ability to record and spy on users. This might be tolerated on small scales within research laboratories but is realistically unacceptable outside of these environments. Indeed the deployment of any static sensors covering a space completely (as any monitoring system must) fosters distrust in a typical environment.

With static sensor deployments seemingly impossible without the simultaneous introduction of user acceptance issues, a more dynamic sensor platforms may be considered: the system users themselves. The goal is to use the information that flows through a positioning system when it is tracking specific objects to infer information implicitly about other (untracked) objects by estimating the current consistency between the physical world and its computational representation.

The solution presented here fundamentally separates the process of monitoring the synchronisation between the physical world and the model from the process of actively updating it to reflect differences. The goal is to collate those signals that propagate through the environment due to the location system itself and examine them for consistency with those signals that would be expected if the current world model was correct.

For this purpose very high accuracy results not necessarily required—it is sufficient to identify gross spatial regions of low consistency and either reduce confidence in inferences made in those areas or target them for updating. In effect, the goal is to get the monitoring gratis, without deploying specialist sensors into the environment other than those used in the tracking solution.

The remainder of this paper introduces a framework for monitoring consistency, demonstrates its application to a working positioning system, and suggests a specific way in which the task of updating inconsistent areas might be tackled.

2   Monitoring Consistency

Figure 1 illustrates a framework that utilises robots (treated here simply as mobile sensor platforms) for the updating of the model. The motions of users produce raw sensor data which is passed to a series of *filters* tailored to the location system in use. The goal of a filter is to combine data from the location system and the current world model to create a snapshot of the consistency between the physical and virtual worlds.

Each filter divides the area it governs into a regular two-dimensional spatial grid (for simplicity only two-dimensional grids are considered, although the ideas are easily extended). Analysis of incoming positioning data is performed with reference to the world model and each grid cell is assigned a value representative of the degree of *spatial consistency*—a measure of the consistency between the physical world and the model associated with it (see Section 2.3). The grids from each filter are then fused to produce an overall spatial consistency grid which is input, in this example, to the control algorithm for the deployed robots. This paper concentrates primarily on the monitoring stage, demonstrating workable filters and a process for forming spatial consistency grids.

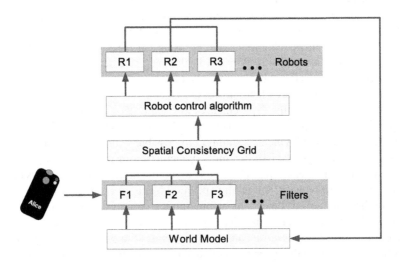

Fig. 1. The process of monitoring and updating

2.1 Classifying Location Systems

Location systems can be classified according to the level of positional data they export. *Generic* location systems offer only a co-ordinate (x,y,z) estimate (and possibly an indication of error or orientation). This information defines the lowest common denominator of data exported by competing location technologies and hence the natural interface on which to standardise if location systems are to become interchangeable components of larger location-aware spaces. Creating a filter to make spatial inferences from generic locations is not easy since much of the data useful for such inferences is peripheral to positioning and thus not represented by generic location data. Nonetheless, generic location systems remain important since commercial location products seem likely to prefer system optimisation over the export of detailed data.

Raw location systems export sensor data along with the co-ordinate result. For example, a system based on time-of-flight measurements might export the raw timings and beacon locations alongside the derived position, or a radio system based on signal strength may offer the signal magnitudes each base station received. The greater volume of data associated with raw systems generally improves the combined output of the relevant filters, although this may come at a cost of increased filter complexity and specialisation.

2.2 Forming Filters

A filter must be tailored to the data it operates on, and may be system dependent. Each filter must convert the incoming data and the computational world model into a common form for comparison; referred to as the *comparison representation*. An example may be a simple occupancy grid, where the volume is segmented into a regular grid used to quantise the world model and separately the data processing output—comparison is then a simple cell-by-cell comparison. In general the comparison representation does not have to be the same for different filters, but should be a natural representation of the filter output that can be easily derived from the current world model.

Two types of filter are defined, according to the possible states that can be determined in its output:

Unambiguous. Given a stream of user positions, an unambiguous filter can assign a per-cell state from the triplet {occupied, empty, undetermined}.
Ambiguous. An ambiguous filter cannot distinguish between all possible states. Commonly, the distinction between a cell that is occupied by an object and a cell that is in the undetermined state cannot be made.

As a concrete example, a filter could be based directly on observed occupancy: if a user is sighted within a cell, that cell must be unoccupied with regards to untracked objects (ignoring random noise and error). This method is able to assert that cells containing a sighting are empty, but can do little to infer the state of the remaining cells that contain no sightings—were they avoided because they were occupied, or are they empty and were not entered by chance? This, then, forms an ambiguous filter. If, however, control was exerted on the sighting distribution, perhaps using a robot that moved to cover all space available (unlike a human, who will typically move to achieve specific goals), the filter may be able to distinguish occupied from undetermined, and the filter would be unambiguous.

2.3 Spatial Consistency

In combining the results of several filters to obtain consistency estimates there is a need to formalise the notion of spatial consistency (SC). SC is defined as the logarithm of the likelihood of being consistent:

$$SC = log_e \left(\frac{P(C)}{P(\bar{C})} \right), \qquad (1)$$

where $P(C)$ represents the probability of consistency and $P(\bar{C})$ the probability of inconsistency. A filter designer may choose to model $P(C)$ and $P(\bar{C})$ directly or model the SC value directly. Many parallels can be drawn with autonomous navigation and mapping, a relatively mature research field that uses similar grid-based methods and likelihood modelling to allow a suitably equipped mobile robot to learn about its environment [9, 10, 11]. Here, however, no control is exerted over the motion of the mobile sensors (the users) and there is a need not only to recognise that spaces are occupied but also to classify or identify the object(s) occupying the space.

The work herein models SC values directly, rather than evaluating $P(C)$ and $P(\bar{C})$ separately. The important requirement is for multiple filters to produce consistent scaling—i.e. an SC value of -0.75 must represent the same consistency across all filters for any comparison to be meaningful. Unfortunately, this is extremely difficult to achieve with heterogeneous filters based on different data. The designer of a filter must therefore strive to approximate the SC value so that multiple filters associate similar SC values to similar consistency states. To aide in this, it is useful to depart from standard probability theory and define a range within which the SC must lie by assigning SC=1 as absolute consistency , SC=0 as undetermined consistency, and SC=-1 as absolute inconsistency. Filter designers then adapt their filter output to produce sensible results on this scale.

When multiple SC values are to be fused, the result must again lie in the range of -1 to 1. To achieve this it is possible to use the mean across *all previously combined SC values*. i.e. Combining two SC values, A and B, may give C=(A+B)/2, and then incorporating D gives E=(A+B+D)/3. Filters that contribute SC values of exactly 0.0 are treated as non-operational and not included in any average.

The following section applies these ideas to a specific location system and presents experimental validation.

3 Case Study: The Bat System

The Bat system was initially described by Ward [12] and developed at AT&T research laboratories at Cambridge, UK. It uses a series of ceiling-mounted ultrasonic receivers installed in a matrix configuration.and with a density of approximately $0.25m^{-2}$. Users wear Bats which can emit ultrasound when so instructed over a radio channel. The radio channel is necessary for accurate time synchronisation (relative to ultrasound, radio waves can be considered to have instantaneous propagation). A command for a particular Bat to emit ultrasound is sent simultaneously to both the receivers and over the radio channel. The receivers measure the elapsed time before they receive an ultrasonic pulse, providing time-of-flight values with which to calculate position estimates. Position is then calculated by converting the measured times to distances using the known speed of ultrasound in air and using a multilateration algorithm. Obtaining a three-dimensional position and an estimate of standard error requires that at least four receivers report reliable measurements: typically eight or nine reliable measurements are reported per-sighting, introducing sufficient redundancy into the positioning algorithm to produce position estimates with an error of approximately 3cm 95% of the time.

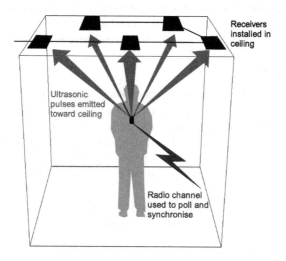

Fig. 2. The Bat system

The Bat system was initially designed to function as a generic location system to assist decoupling of location determination from the dissemination of location data through middleware at a higher level. It has now been adapted to function as a raw system, providing detailed time-of-flight data alongside generic location data. As regards inference about the consistency of world model state, four filters have been developed:

- A linkage filter (generic, ambiguous).
- A journey filter (generic, unambiguous).
- A ray-tracing filter (raw, unambiguous).
- A reflections filter (raw, unambiguous).

These filters are described in detail presently, but it is useful at this juncture to highlight an issue which hinders performance of some filters: the optimisation of the Bat system as a positioning system rather than as a context-gathering tool. As will be seen, much of the consistency state that can be derived from the Bat system depends on the wide propagation of positioning signals. However, the system is designed to *minimise* the extent of propagation of such signals into the environment—receivers are placed in the ceiling and Bats worn at chest height with ultrasonic emitters directed upwards to minimise the likelihood of a signal encountering an obstacle (and reflecting or otherwise interacting such as to reduce the positioning value of the pulse). The result is that signals primarily propagate from chest to ceiling, providing little opportunity to interact with modelled objects (which are typically below chest height) to check consistency.

The ideal solution to this issue is to distribute receivers throughout a space and to redesign the Bats to emit ultrasound more homogeneously. Unfortunately, physical design constraints prevent the system being modified in this way. Instead the filters have been evaluated primarily by maintaining Bats at a lower height than they are designed to be worn at, thereby increasing the chance of signal-object interactions. It should be noted, however, that the creation of positioning systems that promote signal propagation throughout a volume is to be preferred. This can be achieved either by more homogeneous signal emission or by a more uniform distribution of receivers in space.

3.1   The Linkage Filter

The linkage filter [13] attempts to derive consistency information from historical observations of user movements. It monitors each user and when two consecutive sightings for a user are available it quantises the path between the two sighting positions into a series of cell links (Figure 3). Each link is then recorded by incrementing a counter on a linkage grid in memory. Cells with more links to neighbours are assigned a high probability of being unoccupied, whilst cells without links remain undetermined, making the filter ambiguous in nature.

In order to obtain a reliable linkage representation of the environment it is necessary to introduce both spatial and temporal constraints on the incoming location data to account for random errors, low location update rates and unwelcome trends in user

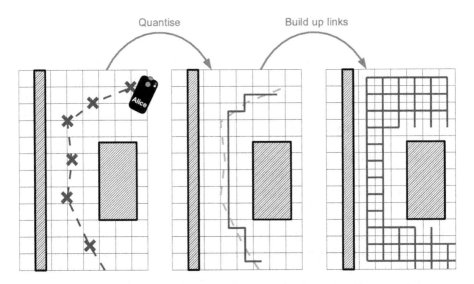

Fig. 3. Linkage filter operation

behaviour. A temporal constraint requires that consecutive sightings must not be separated by more than t_{max}, a time chosen to minimise the chance that people can navigate around large-scale obstacles without being sighted multiple times along the path. The need for spatial constraints stems from a combination of random error and the typical routines of users. Office-based users spend a large proportion of their day in a sedentary state and this is reflected as an extreme non-uniformity in their spatial sighting density which in turn leads to a very high number of votes for links in the vicinity of working areas. As time passes it becomes difficult to establish a reliable vote threshold above which a link is confidently asserted. Introducing a spatial constraint, whereby consecutive sightings must be separated by at least s_{min}, reduces localised high densities of sightings and links.

3.2   The Journey Filter

The linkage filter provides an estimate of the unoccupied cells which can be compared with the world model, but ambiguity between empty cells and undetermined cells hinders its use. An alternative filter considers entire *journeys* rather than individual cell transitions. A journey is defined as a route taken by a user in traversing from one area of low sighting activity to another. The involvement of humans is important since the interest is in any deviation from the direct 'as the crow flies' path between start and destination, from which the filter aims to construct an occupancy grid. Humans are naturally good at solving path planning problems, minimising how far they must move. The journey filter capitalises on this by analyzing each journey relative to the optimum.

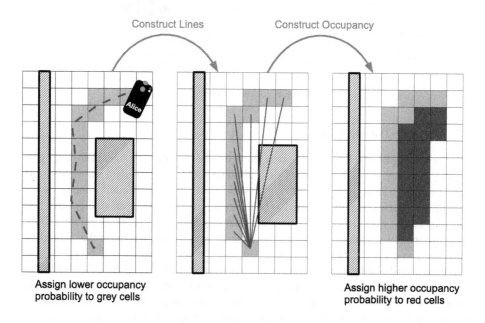

Fig. 4. Journey filter operation

Figure 4 illustrates the analysis algorithm for a room containing a rectangular obstacle. A user, Alice, moves from top to bottom along the broken line shown. This path is quantised onto the grid and each cell involved has its probability of occupancy decreased. Thereafter, a series of sub-journeys are generated by treating each cell along the path as being the end of the journey (the solid red lines of Figure 4 illustrate this). These lines are quantised as before, except that the cells they intersect have their probability of occupancy *increased*, representing the fact that they were avoided, despite being shorter routes.

As the occupied, undetermined and unoccupied states are explicitly estimated, the filter is unambiguous in nature. Figure 5 shows a real-world example derived from a user walking around a table (centre, shown in blue outline) a small number of times. Colour coding is used to represent the SC values; red-white-green (or dark grey-white-light grey for monochrome copies) for occupied-undetermined-unoccupied. The filter is able to clearly distinguish a core of occupied cells as expected. By comparison, usage of the linkage filter would not determine the state of the core cells.

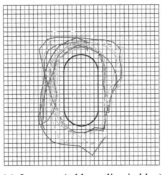
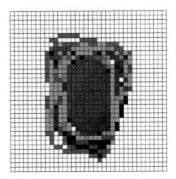

(a) Journeys (table outline in blue)   (b) Occupancy grid (from journey analysis)

Fig. 5. Journey analysis

3.3   The Ray-Tracing Filter

The ray-tracing filter is based on the straightforward notion that if a signal can propagate directly between two known points then all cells that the signal path impinges upon must be empty. Thus simply quantizing all valid paths from the emitter to the receivers determines a series of cells expected to be unoccupied, making this an ambiguous filter. The idea can be extended to make the filter unambiguous by considering not only those signal propagations that were determined valid ('rays'), but also those that would be *expected* given the receiver distribution and the emitter position ('pseudo-rays') [14, 15]. Figure 6 illustrates the use of pseudo-rays in estimating occupancy. First the local receiver distribution and a model of the ultrasonic propagation is used to determine the pseudo-rays, assuming no obstructions. These are quantised onto a three-dimensional grid by incrementing a counter associated with each cell that they impinge upon. A secondary per-cell counter is used to record the number of rays that intersect in a similar

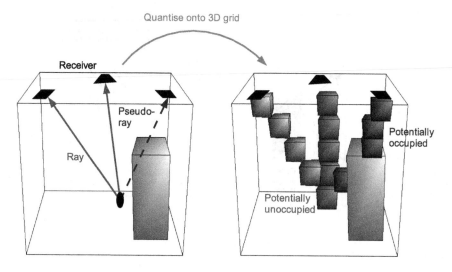

Fig. 6. Ray-tracing

fashion. Following the processing of a series of sightings in an area, a probabilistic occupancy grid can be constructed by considering the number of intersecting pseudo-rays (p) with respect to the number of intersecting rays (r). For example, if $p \approx r$ then the cell is expected to be unoccupied; if $p \gg r$ the cell should be to be occupied, and if $p \approx r \approx 0$ only the undetermined state is applicable. The resulting probabilistic grid forms the comparison representation.

3.4   The Reflections Filter

Experiences with ray-tracing show that it has potential to work well in the centre of rooms, but suffers near the perimeters where receiver visibility is typically limited. To monitor these areas a filter which analyzes signal reflections is more applicable. The analysis is based on a technique known as Single Reflection Spatial Voting (SRSV) [16]. This approach uses the entire set of time-of-flight readings taken for each sighting. Those that are self-consistent are used to determine the Bat position, whilst those that are not are treated as multipathed signals that have undergone a single specular re-flection from source to receiver. For each multipathed signal, an ellipsoidal surface of points from which it could have reflected is constructed (Figure 7). The premise is that, given a statistically significant number of such ellipsoids, the reflecting surface becomes self-evident in the data.

To proceed, the volume is again partitioned into regular cells. With each incoming reflection measurement, a reflection ellipsoid is constructed and quantised onto this grid. Within each cell on the ellipsoid boundary, the local normal vector is computed and associated with the cell. If the cell contains a reflecting surface, the accrued nor-mal vectors should tend to point in the direction of the surface normal vector. Thus an average of the associated normal vectors is taken per-cell, forming a quantised vector

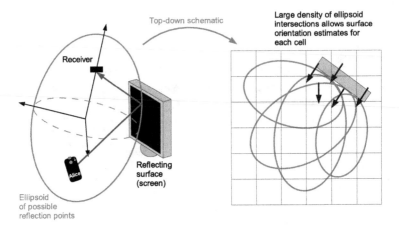

Fig. 7. Reflection

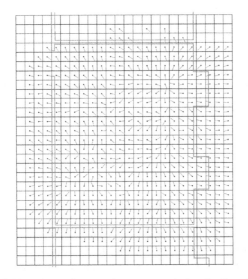

Fig. 8. Experimental comparison representation for the reflection filter

field that is the comparison representation. Figure 8 shows such a representation for an empty room (primary reflecting surfaces being walls). The data was collected across a single day using location data from Bat system users. It is clear that the averaged normals are near-perpendicular to the walls and oriented randomly elsewhere.

The comparison representation can be generated from the base world model by breaking objects down into visible surfaces, and quantizing these surfaces onto a grid. Reflected signals are of use not only at perimeters of physical boundaries but also at the perimeters of user sighting distributions—they have the potential to extend this perimeter since they penetrate into the environment to a greater extent.

3.5   Spatial Consistency Results

Figure 9 is a graphical depiction of an experimental spatial consistency grid based on 31289 positions collected in a single room over approximately 30 minutes using the Bat system in a high quality-of-service mode. Cells are colour-coded using their SC value: the colour map is a linear progression from green (consistent) to white (undetermined) to red (inconsistent) (this progression is light grey-white-dark grey for monochrome copies).

The room was setup with a square table sited in the centre of the room, but the corresponding world model had the table location to be the top left room corner (Figure 9(a)). The consistency grid shows two areas of major inconsistency, corresponding to the presence of an object where the world model claims there is none (central) and the lack of an object where the world model claims there to be one.

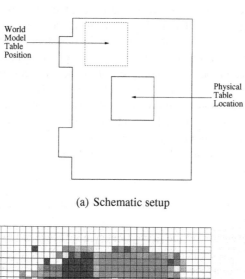

(a) Schematic setup

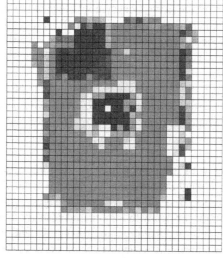

(b) SC grid

Fig. 9. An experimental spatial consistency grid

Importantly, the grid resolution need only approximate the smallest object of interest—this methodology makes no attempt to identify objects at this stage, simply zones of inconsistency. Thus object boundaries do not need to be accurately determined for the purposes of shape recognition. Prioritising the correction of inconsistent regions can then be achieved either through predetermined ranking (i.e. using the world model to infer which semantic entity has been moved and how important it is to capture that change) or by inconsistent area (i.e. a large area of inconsistency should be corrected before a smaller one).

4   Updating the Model

Once a spatial consistency grid has been established, the question of how best to update the model to remove the identified inconsistencies remains. In order to gain high accuracy, an array of sensors in the locality of the inconsistency is required, and yet we do not wish to deploy such sensors statically to avoid privacy implications. At present it seems unlikely that a human can be removed from the feedback loop associated with updating a model. However, a mobile sensor platform could be used to investigate an area using an array of sensors and relay the results back to a human for assistance. For example, such a platform may be able to determine the position and orientation of an object, but require a human to identify the object; this could be done quickly and easily using a photograph. The use of such automatons addresses the issue of user privacy and trust in a series of ways:

– Users might assign spatial zones and times within which robot presence is prohibited or allowed.
– The presence of a robot provides a physical symbol of when sensors are active in an area (the physical aspect is important; a static camera set to stop capturing during certain periods still has the ability to continue to capture covertly, and users of the space would thus never be certain of whether they were being monitored).

An alternative approach is to label objects in such a way as to be machine readable. Augmenting objects in the world with unobtrusive RFID tags is one approach investigated [17, 18, 19]. A robot would then carry a reader and hence have an idea of those objects it is expecting to sense in its surroundings, allowing for model-based recognition techniques.

The use of robots carries with it a compromise in the form of an increase in the latency of updates. Clearly a single robot cannot monitor a large space and hope to provide an instantaneous response to all environment changes. A plurality of robots would solve this problem but this brings with it increased costs (both of a recurrent and non-recurrent nature), increased user irritation and disruption, and decreased privacy (the ideal for low latency monitoring is to deploy a high density of robots, but this is in effect a reversion to the use of static pervasive sensors). In this methodology, once a spatial consistency grid is established it is used to prioritise the motions of a small team of robots which must accurately survey each area in a timely manner.

5 Conclusions

This paper has identified a need to maintain world models within high-accuracy location-aware systems. It has introduced a novel approach that separates monitoring consistency between physical and virtual worlds from correcting it, presenting a metric and a framework to do so. We have implemented the monitoring framework using a live positioning system and shown it to function well. Results to date suggest that the separation of monitoring from updating in the life cycle of a world model is a useful approach in real world scenarios. Knowledge that a zone is not consistent is arguably of more use than the approach of correcting mistakes as they are found, which necessitates a dense sensor network (static or dynamic) to be present before action can be taken.

A number of approaches to correcting the model after areas of inconsistency are apparent have been presented and there is much scope for future work.

References

1. R. Want, A. Hopper, V. Falcao, and J. Gibbons. The Active Badge Location System. *ACM Transactions on Information Systems*, January 1992.
2. Roy Want, Bill Schilit, Norman Adams, Rich Gold, Karin Petersen, John Ellis, David Goldberg, and Mark Weiser. The PARCTAB Ubiquitous Computing Experiment. Technical Report CSL-95-1, Xerox Palo Alto Research Center, March 1995.
3. M. Addlesee, R. Curwen, S. Hodges, J. Newman, P. Steggles, A. Ward, and A. Hopper. Implementing a sentient computing system. *IEEE Computer*, 34(8), August 2001.
4. T. Richardson, Q. Stafford-Fraser, K. R. Wood, and A. Hopper. Virtual Network Computing. *IEEE Internet Computing*, 2(1), January 1998.
5. A. Bobick, S. Intille, J. Davis, F. Baird, C. Pinhanez, L. Campbell, Y. Ivanov, A. Schutte, and A. Wilson. The KidsRoom: A Perceptually-Based Interactive and Immersive Story Environment. Technical Report 398, MIT Media Lab, December 1996.
6. J. Krumm, S. Harris, B. Meyers, B Brumitt, M. Hale, and S.. Shafer. Multi-Camera Multi-Person Tracking for EasyLiving. In *Proceedings of the Third International Workshop on Visual Surveillance*, July 2000.
7. H. Kato, M. Billinghurst, I. Poupyrev, K. Imamoto, and K. Tachibana. Virtual Object Manipulation on a Table-Top AR Environment. In *Proceedings of ISAR 2000*, October 2000.
8. Jun Rekimoto. Matrix: A realtime object identification and registration method for augmented reality. In *Proceedings of Asia Pacific Computer Human Interaction*, pages 63–68, July 1998.
9. H. P. Moravec. Sensor fusion in certainty grids for mobile robots. *AI Magazine*, 9:61–74, 1988.
10. M.C. Martin and H.P. Moravec. Robot Evidence Grids. Technical Report CMU-RI-TR-96-06, Carnegie Mellon University, 1996.
11. S. Thrun. Learning maps for indoor mobile robot navigation. *Artificial Intelligence*, 99(1):21–71, 1998.
12. A. M. R. Ward. *Sensor-driven Computing*. PhD thesis, Cambridge University, August 1998.
13. R.K. Harle and A. Hopper. Using Personnel Movements For Indoor Autonomous Environment Discovery. In *Proceedings of the First IEEE International Conference on Pervasive Computing and Communications, Fort Worth, TX, US (PerCom 2003)*, pages 125–132, March 2003.

14. R. K. Harle and A. Hopper. Building World Models By Ray-tracing Within Ceiling-Mounted Positioning Systems. In *Proceedings of UbiComp, Seattle, Washington, US*, October 2003.

15. R. K. Harle and A. Hopper. Dynamic World Models from Ray-tracing. In *Proceedings of the Second IEEE International Conference on Pervasive Computing and Communications, Orlando, Florida (PerCom 2004)*, March 2004.

16. R. K. Harle, A. Ward, and A. Hopper. A Novel Method for Discovering Reflective Surfaces Using Indoor Positioning Systems. In *Proceedings of the First International Conference on Mobile Systems, Applications, and Services, San Francisco, California, US (Mobisys 2003)*, 2003.

17. G. Borriello, W. Brunette, M. Hall, C. Hartung, and C. Tangney. Reminding adbout Tagged Objects using Passive RFIDs. In *Proceedings of the 6th International Conference on Ubiquitous Computing (UbiComp 2004), Nottingham, UK*, September 2004.

18. D. Hahnel, W. Burgand, D. Fox, K. Fishkin, and M. Philipose. Mapping and Localization with RFID Technology. In *Proceedings of the 2004 IEEE International Conference on Robotics and Automation (ICRA '04)*, 2004.

19. J. Brusey, C. Floerkemeier, M. Harrison, and M. Fletcher. Reasoning about Uncertainty in Location Identification with RFID. In *Workshop on Reasoning with Uncertainty in Robotics*, August 2003.

Shadow: A Middleware in Pervasive Computing Environment for User Controllable Privacy Protection

Wentian Lu, Jun Li, Xianping Tao, Xiaoxing Ma, and Jian Lu

State Key Laboratory for Novel Software Technology
Nanjing University, 210093 Nanjing, China
{luwt, lijun, txp, xxm, lj}@ics.nju.edu.cn

Abstract. In ubiquitous and pervasive computing, after data owner's information is collected, data collector should be careful of disclosing data owner's information for privacy reasons. In this paper, we present requirements and challenges when designing solutions for such data collector end protection. Policies, accuracy and anonymity of context should be all taken into account. Based on this, we design a middleware Shadow for user controllable privacy protection, which is deployed on data collectors who have large volume of data and powerful computation abilities. Shadow has a contextual rule based access control policy mechanism, enriched with methods of generating blurred context and guaranteeing information anonymous, and we implement it under an ontology based context model.

1 Introduction

Pervasive computing is to enhance the environment by embedding many computers that are gracefully integrated with human users. Actually it also offers opportunities for new privacy risks and people represent their fear [1]. So, privacy-sensitive or privacy-aware needs for pervasive computing have risen.

Generally, a pervasive environment contains three main participants, data owner, data collector and data user as showed in Figure 1. Data owners are the source of context information. When we focus on privacy issues, persons are the main data owners. Data collector includes sensing hardware,storage databases and relevant software. They use sensors or devices to gather information of data owner such as location, store and manage them. Data users are those who obtain context information of data owner through data collector for specific usage. They usually are persons and pervasive applications.

When a data owner wants to preserve his privacy in pervasive environment, he should conceivably decide to whom and how his data should be released. On one hand, privacy may be revealed in interactions between data owner and data collector(dashed line in Figure 1). To protect such disclosure, it requires the user to carry a device that is compatible with beacons and powerful enough to make access control and encryption algorithm, or to simply shut down the devices or throw away smart tags. On the other hand, after context is gathered by data

P. Havinga et al. (Eds.): EUROSSC 2006, LNCS 4272, pp. 143–158, 2006.

collector, we also should be responsible for careful information disclosure to data user(solid line), because unappropriate released information will be dangerous to data owner's privacy. Actually nowadays not every one has a operative device with him. Even though he has, these devices usually lie on low computation power which make them difficult to execute complicated algorithms. In addition, most time our information are more or less stored on a data collector. Therefore strengthening data collector end protection for data owner is important.

Fig. 1. Participants in Pervasive Environment

In this paper, we focus the privacy problem which data collector should take into consideration. We assume that data collector is trustable and can precisely get context information from data owner. There are two main differences between data owner and data collector. The latter is more powerful in computation so we can take some complicated methods. And it owns a large amount of data over different data owners, which can be used to generate more effective control. Both of these are what couldn't be done in data owner end protection.

To achieve data collector end protection, we believe a middleware approach is a solution, considering different data users and interactions between queries. Our middleware *Shadow* is based on a pervasive computing environment FollowMe [2]. The contributions is threefold. First, we discuss challenges that arise when specifying privacy control policy in a pervasive environment. Second, we design a contextual rule based access control policy mechanism enriched with methods of generating blurred context and guaranteeing information anonymous. Third, we present our implementation with an ontology based context model and evaluation of the middleware.

The structure of the paper is as follow: Section 2 gives out a motivated scenario and general requirements. In section 3 we analyze context and its accuracy. In section 4, challenges for policy design are discussed, followed by detailed design of system including context modeling, anonymity generation and contextual rule based access control in section 5. Section 6 provides information of entire architecture, implementation, evaluation and discussion. And then related work is presented in section 7. We end with conclusion and future work in section 8.

2   Motivation and General Requirements

The following scenario illustrates the requirements for privacy needs in a context-aware environment.

Tom is a graduate student in computer science department. His research interest is in P2P network and he is working with network group. There is a large-scale pervasive and ubiquitous computing infrastructure which can obtain his location, activity or other information about him through different technologies, such as GPS, sensor, smart tag and mobile device.

Now Tom's advisor wants to know where he is for some arrangement of academic discussion. However, Tom thinks this tracking should be happened only during work hours and when he is in the area of working place. If Tom wants to track his advisor's location, he may be only get a low accurate information, such as his advisor is in builing1 but not a specific room in builing1 or detailed coordinates.

During the working day, when an application query the persons in one room where Tom is, he just keeps his identity anonymously by returning a anonym generated by randomization or replying as a member of network group.

Tom also subscribes a reminder service, which will remind him what should do on the schedule when he is at the right place and right time. He only wants to provide one pseudo-name during different periods of time and this should enable the service to obtain Tom's information without true identifier disclosed from data collector in which data is organized by true name.

This scenario also illustrates the richness of constraints(or preferences) data owners might want to apply to control the distribution of their information. In our examples, Tom restricts access to his information in several ways:

- *Inquirer.* In most cases, Tom restricts access to his information to specific inquirers(such as his advisor, his friend or some specific applications).
- *Accuracy.* Tom confines his information disclosure with different accuracy. He might want to tell someone he is in campus but for others he will tell them the precise location.
- *Contextual Condition.* Disclosure of information is relevant to contextual conditions such as time and location. During the work time, or on weekends are different. At home or in the office are also should be taken into consideration.
- *Anonymity.* Tom restricts access to his information by releasing a fake name to hide his true identity.

3   Context and Its Accuracy

In pervasive environment, any context information related to persons contains two aspects: who it belongs to and what it describes. Take "Tom is at home" as an example. The identity of data is the person Tom, which may connect to a full name or a unique social number. The content of this context is location "at home". Anyone who wants to use context should take both aspects into consideration. So do privacy problems. Identity refers to who you are and context reveals what you do, where you are or other things like this.

Identity is data about who you are. Strictly, it is a identifier in context query e.g. a person's name or a student ID. For privacy concerns, this identifier could be anonymously representative. We used to think anonymity is the privacy of identity. We can divide anonymity into two cases: persistent anonymity, where user maintains a persistent online fake name which is not connected with the user's physical identity("true name"), and one-time anonymity, where an online faked name lasts for just once use. More generally, identity includes personal data which could help to identify a person such as gender, age, occupation, address and telephone number.

Context data such as location, activity and schedule are different from traditional personal information because of their time-relevant attributes. They are transient compared to those information like occupation.

Generally, we believe a context can be viewed as [identity, location, activity, time]. Identity may be identifiers of persons or personal information which will identify someone. So identity is a multi-dimensional value. Location and activity refer to where the person is and what he is doing. Time represents when this context is observed.

Snekkenes [3] used lattice to define accuracy of context, including identity, location. Here we adapt that definition and Figure 2 shows one concrete example of accuracy structure of a context. Actually when combining different accurate values of each field, we can obtain types of context generated from the same original context, like "Tom is attending a seminar about 'Chord' in Room1". For privacy concerns, the disclosed information is always be selected in the accuracy structure according to different policies. That means policy decides what accuracy is necessary and privacy is protected by policies.

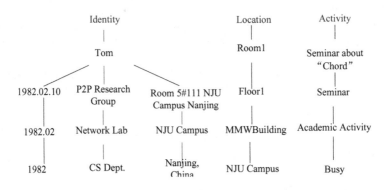

Fig. 2. Accuracy Structure of A Context

4   Policy Issues

Before we take deep considerations of how to make policy for privacy, we should make a clear understanding of what the context will be queried by data users.

- *Person Based Query*–this type of query are such questions: Where are you? What are you doing? The data user needs to know the location, activity or other information of a specific person, who is identified by a identifier. More specifically, it could be divided into two subtypes: *Single Based Query* and *Group Based Query*. The former uses true identifier or pseudonym. The latter uses group identifiers such as P2P research group in Figure 2, which means to query all members of that specific group.

– *Attribute Based Query*–this type of query are such questions: Who are in that room? Who are having the seminar? Data user wants to get lists of persons at specific location or doing a specific thing, or other concrete attributes.

G.Myles [4] divided query into three categories and additional one is asynchronous requests, which is a kind of "event" information, such as when users enter or leave specific areas. However, it can be merged into our first category. There are no basically differences between them.

4.1 Single Policy vs. Multi Policies

Hengartner [5] proposed user policy and room policy which are corresponding to our two types of queries. However, these multi policies will cause some problems and we believe a single policy–owner policy will be enough for the privacy needs.

– The room policy is restricted in room situation and our attribute based query is not only to get room-level location information. So it is difficult to establish different policies for all kinds of attributes such as activity and floor-level location.
– Multi policies will cause conflicts. For example, assume that Tom does not Bob to locate him but the room policy agrees. More complicated the multi policies are, more chances of conflicting will happen.
– Using only owner policy will also satisfy attribute based query. Assume someone queries a room and wants to know who are there. Before return the person list, system could check the owner policy to decide what should be returned, a anonym, a obfuscated identity such as a member of P2P research group.

4.2 Anonymity of Identifier and Context

Producing anonymous context information is nontrivial task. When we use network communication to get information and hardware data or network protocol, it will reveal your identity, such as MAC and IP address. To address this type of issues, Crowds [6], Onion Routing [7], or mix Zones [8] provides anonymity in communication. Here we focus on anonymous context queried by data user and anonymity in communication can achieved by technologies described.

Anonymity of identifier will happen when attribute based query initiates. When system receives a location query that requires it to reveal a owner's identifier, his policy should be checked and system determines whether to return the owner's pseudonym, anonym or true identifier.

Anonymity of context is more complicated because applications might be able to deduce the mapping between temporary and permanent identifiers by observing movement patterns. For example, you disclose locations to a query with anonym. However, the data user find all locations are identical for the same time period. Thus it may infer that they belongs to the same person in spite of different randomized anonyms, which can be called matching inference attack described in [9]. For the purpose of being unidentifiable, Sweeney have proposed the concept of $k-$anonymity [10]. A $k-$anonymous released dataset has the property that each record is indistinguishable from at least other $k-1$ records within the dataset. The larger the value of k, the greater the implied privacy since no individual can be identified with probability exceeding $1/k$ through linking attacks alone. We believe this is a useful way keep context anonymous.

4.3   Personal Policy and Impersonal Guarantee

Privacy is a terminology with *personal* and *impersonal* aspects. Since different persons have different views on the same attribute, we say it is personal. Women may feel age is a more sensitive information than men do. Thus, before women decide to tell their ages, they treat it as an important privacy concern. For men, they may treat age as normal information and can tell others easily. This is the basic reason why data owners should define their own policies. And our policy mechanism should provide such abilities for data owners to customize their personal policy definitions.

Impersonal guarantee is relevant to anonymity of context. People don't want to be identified by those attribute based queries so they use anonymous identifiers or blurred context. But they do not know whether the information he provides is safe from inference attack, which will help malicious data users to link different information to the same person or disclose more accurate personal data. Since this can not be achieved by policies or be realized before query happens, system should take the responsibility for anonymity check.

5   System Design

Based on our discussion in previous section, we now present the design of our middleware to support privacy control with policy in a pervasive environment. We build on three main parts. First, ontology based model for context building and accuracy generating. Second, how to achieve anonymity of context. Third, contextual rule based access control policy.

5.1   Ontology Based Context Model and Context Blurring

An ontology defines a common vocabulary for researchers who need to share information in domains. Its advantages includes knowledge sharing, easy reuse of knowledge and enabled knowledge analyzing. Here ontology based context model provides formal representation and formal logic inference with a semantic meaning for each context data. Our context model consists of 2 parts: ontology and its instance. The ontology is a set of shared vocabularies of concepts and the interrelationships among these concepts. And in our work, we choose OWL as ontology description language in which concept refers to class and interrelationship refers to property. Instances of ontology include both persistent contexts and dynamic contexts. Persistent contexts usually have a long life period and they would be combined with dynamic contexts during reasoning process. We also use triples to describe context in the form of (subject, predicate, object) according to RDF. For example, the context "Tom is a student" is modeled as (Tom, rdf:type,student). Dynamic contexts with transient characteristics only have a short lifetime in the system, such as "Tom is in Room1". As Figure 3 shows, every person has some attributes such as location and occupation. And every attribute concept (e.g. location) has its own hierarchial definition. Figure 4 represents the part of persistent context of location in our system and they will be used in context reasoning.

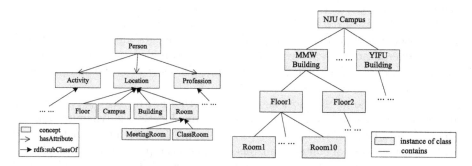

Fig. 3. Part of Our Ontology Model   **Fig. 4.** Part of Persistent Context Structure for Location

Activity is some kind of complex because it can not be simply sensed by sensors like location information. To monitor on person's schedule is one approach, but it could not handle activities which can not be found in schedule. With ontology reasoner and rule reasoner, we can generate high-level activity context. This is one important reason we use ontology as our context model. Here are two rules used to generate talk activity.

TalkRule : $(?x \quad locateIn \quad ?room), (?y \quad locateIn \quad ?room), (?room \quad rdf:$ $type \quad Room), (?x \quad sound \quad high)- > (?x \quad talkWith \quad ?y)$

Generating Blurred Context

Blurring is usually based on a generalization. One important problem is to decide what a value is generalized to. Recently some researchers tried to use a data mining method to split the value space [11]. They utilize a data mining method and we know that may be precise to get a detailed generalization but sometimes it is heavy burden for system and we do not need this anonymity all the time. We believe ontology can be viewed some kind of hierarchial structure in which higher class usually have a more general and blurred meaning than lower class but always keep sematic consistence. Therefore, ontology model and its reasoning ability can be used to blur context.

Ontology directed blurring is based on two kinds of inference: superclass inference, and TransitiveProperty inference. Super class rule means an instance of class A is also the instance of A's superclass. A student specialized in "P2P Network" can be generalized to "Network Research", because the latter is superclass of the former. TransitiveProperty rule means if $a \rightarrow b$ and $b \rightarrow c$, we get $a \rightarrow c$. For example, if we have a precise data $(Tom, locatedIn, Room309)$ and a persistent context $(Room309, locatedIn, MMWBuilding)$, a new data is generated $(Tom, locatedIn, MMWBuilding)$ which conveys little information. Accuracy level structure of Figure 2 is generated in this way.

5.2   Achieving $k-$Anonymous Context

$K-$anonymity can promise your information unidentified without other $k - 1$ datasets. This type of anonymity will be a effective way to keep your information hiding in other $k - 1$ datasets which means even though you are anonymous in a $k-$sized set. For typical personal information like age, occupation and birthday can be generalized to $k-$anonymous data using methods introduced by DataFly system [12]. However, mechanism for dynamic pervasive data should be different from that of static database tables.

Location is one of the most significant information in most pervasive environment and it changes very quickly. That means location is dynamic, nonstop-changing and time relevant. More specifically, firstly, blurring obsolete context will destroy validity of context and generate potential privacy risk. Secondly, information disclosure without attention to history will help others to link current and previous data to identify person easily.

We use a spatial tree like Figure 4, which is constructed by persistent context in our system and relationship among nodes are maintained according to ontology definition of location. We would like to mention this kind of structure can be extended to other attributes like activity. In our algorithm, every node of the spatial tree records the current k value, and S_{id} stores all levels of precision location context for one person. When a new context comes in, the algorithm will judge if it is a new position. If it is, the tree will adjust all the k values of nodes and S_{id} will be replaced. Easily we can find father node usually will have higher k than child nodes because it gathers more generalized context from them. Here we have a trick when a query comes and no enough k value are stratified right away, the system will wait for a small period of time to wait for larger k. That means when more people enters the area, it is more possible to get a suitable k.

5.3   Contextual Rule Based Access Control with Identifier Manager

A simple access control is not enough to handle privacy needs, especially in a context-aware environment. We need to control contextual conditions like location and time which is showed in the scenario. Moreover, reducing context accuracy and control degree of anonymity should be taken into consideration too. To achieve these purposes, we develop an enhanced access control with contextual rules according to policy issues discussed in section 4.

As Figure 5 shows, *Policy* is defined by *Owner*, and is set to a *Recipient*. Every policy has some *Context Types* allowed to disclose, such as location and activity. To each allowed context type, its disclosure is controlled by some *Rules*. Each rule is defined by three properties. *Condition* describes under what condition, this rule is triggered. *Accuracy* tells the engine what level of accuracy should context data be given to this rule. It determines how much the context will be

---

Algorithm 1: Generating $k-$anonymity context

Data: A tree structure representing space
Result: $k-$anonymity status for context
begin

    S_{id} contains all location context of one person who is represented by id

    for *all nodes of tree* **do**
        └ $node.k \longleftarrow 0$

    while *new precise location context* $(id, newL, newT)$ *comes in* **do**
        $//(id, newL, newT)$ represents context $(Identity, Location, Time)$
        if S_{id} *does not contain location* $newL$ **then**
            $p \longleftarrow node_1(node_1$'s location $== newL)$
            $q \longleftarrow node_2(node_2$'s location $== L \wedge L$ is precise location in $S_{id})$
            repeat
                $p.k \longleftarrow p.k + 1 \quad q.k \longleftarrow q.k - 1$
                $p \longleftarrow p.fatherNode \quad q \longleftarrow q.fatherNode$
            until $p == q$
            generate blurred context with ontology model and persistent context
            $\{(id, L_m, newT)\}, m \in \mathbb{N}, L_0 = newT$
            $S_{id} \longleftarrow \{\}$
            for *all* $L_n, n \in \mathbb{Z}$ **do**
                $k \longleftarrow tmpnode.k(tmpnode$'s location $== L_n)$
                └ $S_{id} \longleftarrow S_{id} \bigcup \{(id, L_n, newT, k)\}$

end

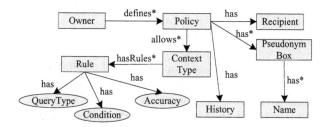

Fig. 5. Ontology Concepts for Policy

generalized or can be described by k in $k-$anonymity. *Query Type* tells system what the query is, person based or attribute based, true name based or anonymous identifier based. Query type also determines what kind of information should be return. If query is person based, return is the attribute. Oppositely, attribute based will return identifier. For returning a persistent-anonymous identifier, pseudonym will pick up the current used name in *PseudonymBox*. And *PseudonymBox* records the used names and current using pseudonym. *History* records previously disclosed context. This is important because through history, people knows when and what has been disclosed to a recipient, under what accuracy and using what degree of anonymity.

```
<policy rdf:about="&shadow;advisor">
  <Recipient rdf:about="&shadow;Jack"/>
  <contexttype rdf:resource="&shadow;Location">
    <rule>
      <querytype>
      person-based-query
      </querytype>
      <condition>
      time=9 a.m. to 5 p.m.
      </condition>
      <accuracy>precise</accuracy>
    </rule>
  </contexttype>
</policy>
```

(a) Policy 1

```
<policy rdf:about="&shadow;custom1">
  <Recipient rdf:about="&shadow;RoomMonitorService"/>
  <contexttype rdf:resource="&shadow;Location">
    <rule>
      <querytype>attribute-based-query</querytype>
      <condition>
      time=9 a.m. to 5 p.m.
      && owner.activity=academic-activity
      </condition>
      <accuracy>P2P Research Group</accuracy>
    </rule>
  </contexttype>
</policy>
```

(b) Policy 2

```
<policy rdf:about="&shadow;custom2">
  <Recipient rdf:about="&shadow;YouAreHereMapService"/>
  <contexttype rdf:resource="&shadow;Location">
    <rule>
      <querytype>pseudonym-based-query</querytype>
      <condition>owner.location=MMWBuilding</condition>
      <accuracy>precise</accuracy>
    </rule>
    <rule>
      <querytype>pseudonym-based-query</querytype>
      <condition>owner.location!=MMWBuilding</condition>
      <accuracy>6-anonymity</accuracy>
    </rule>
  </contexttype>
</policy>
```

(c) Policy 3

Fig. 6. Policy examples of Tom

Figure 6 illustrates three policies Tom has made for different data users. Policy 1 6(a) represents "if my advisor Jack wants to locates me during working time, I will let him know my precise location." Policy 2 6(b) tells us "if Room-Monitor service wants to get the people list in a specific room and I am in that room, I will let it know I am a member of P2P research group only when I was having academic activity such as seminar or lecture during my working time ". Policy 3 6(c) means "I will use a pseudonym with 'You are here' map service. If I am in MMWBuilding I'll let it know my precise location because the structure of the building is complex and precise location is useful. But if I am out of that building, I would provide blurred location, and this time the system should promise this location is 6-anonymity".

Pseudonym Changing and Matching

The advantages of using pseudonym are keeping anonymity and meanwhile enjoying abilities of personalization. Because this is a kind of persistent anonymity, it is more vulnerable to be identified than one-time anonymity. So we should change pseudonym when there is a high possibility to identify the specific person. This trigger is based on disclosure history of a policy. Data Owners can set some thresholds to control data disclosure e.g. the maximum size of disclosure set. Or if some very sensitive information is disclosed under this pseudonym, it is necessary to change to a new pseudonym for preventing identifying easily. One finds a pseudonym enter a special room and only three persons are allowed to enter it. At this point, third party service can identify user with a little more knowledge. So this pseudonym should be ended. Data Owners can predefine some new pseudonyms for future usage, or new pseudonym is produced by system through randomization. New selected pseudonym will be the current one and previous one will be pushed into used pseudonym list.

When a data user, may be a pervasive service, tries to query context using pseudonym, system will execute a pseudonym matching process for context fetching. That means finding out the true name of the pseudonym and utilizing it to continue query. Before system returns results to data user, a reversed process is taken. System keeps an index of pseudonyms and make sure their distinctness. The whole process promises data user will get the context with a pseudonym and privacy is also protected for data owner.

6   Middleware *Shadow*

Our privacy middleware is built on top of the previous work, a context-awareness infrastructure FollowMe [2] which is responsible for the gathering and storing of, reasoning on and dissemination of contextual information on the user and the services the user is interacting with. FollowMe is also a pluggable infrastructure which can serve as both a runtime environment and a software library.

6.1   Overview

As Figure 7 shows, query interfaces are for data users, which could be human users or pervasive applications. There are person based query interface and attribute based query interface such as location query and activity query. Before they get into the system, access control engine will check their identity. This process is kind of authentication. After that, if the query is described by pseudonym, identifier manager will match it to true name by an index. Then query could should be filtered by policy definition and triggered by contextual rule engine. After all of this, system will fetch context data as policy needs. The result could be precise information, blurred according to ontology or check by $k-$anonymity algorithm to promise unidentifiability. At last, result returns to data user, if necessary, anonymous identifier will be transformed by identifier manager. User interface is also provided for data owners to customize his own policy definitions.

6.2   Implementation Details

The infrastructure is FollowMe, which is based on OSGi[1] and contains sensors and RFID devices. We build up our middleware Shadow on the top of it. Shadow is implemented in the Java language and client users can access to it through

---

[1] OSGi, Open Service Gateway Initiative: http://www.osgi.org

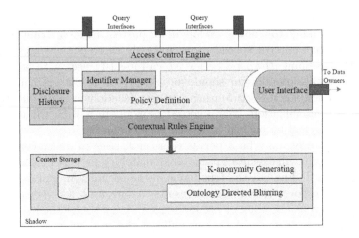

Fig. 7. Architecture of Shadow Middleware

http protocol or Java RMI based applications. For ontology, our implementation is restricted in OWL-Lite, which is equipped with kinds of expressive description such as "TransitiveProperty", "SymmetricProperty". In addition, OWL-Lite decreases the number of instances of ontology compared to OWL-DL and OWL-Full and promise that the reasoners will be relatively simpler than those of other two. For reasoning, our work is built up by Jena API[2] and modified RDQL[3] engine to support different queries.

6.3   Evaluation and Discussion

We do some experiments to evaluate the middleware. Firstly we set up one server and two clients. The central server is also the Linux workstation with 4G RAM and 2 Xeon CPUs. Context is provided the central server at different frequencies while an application is running at another client node and queries contexts from the central server. One burden of the middleware is ontology based reasoning and $k-$anonymity management, we find that when context generates over 40 per minute with ten thousand persistent context, the latency of query will increase greatly and even reaches over 60 seconds. This could be unbearable. Secondly, we try to figure out how many policies should be checked during a time unit. Because every checking will need to parse XML base files, high cost on this is not suitable. The number of checked policies is effected by many factors: the number of involved persons and possible places(we describe this by average k for each location), the possibility of each query type(we use (s, g, a) to delegate single-based, group-based, attribute-based queries respectively) and query frequency(x-axis in Figure 8). We can find that our method is not efficient for too crowded areas such as supermarket. Also when group and attribute based query happens

---

[2] Jena2 Semantic Web Toolkit: http://www.hpl.hp.com/semweb/jena2.htm
[3] RDQL tutorial: http://jena.sourceforge.net/tutorial/RDQL/index.html

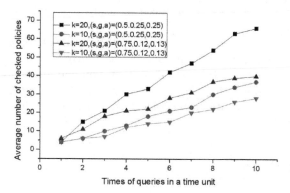

Fig. 8. Average number of checked policies

frequently, the number of checking will increase, which drive us to the way of something like room policy.

In a summary, we get some conclusions. Firstly, this logic inference based context and k–anonymity generating is time consuming. When we have high speed of context stream, it isn't suitable for real-time applications. But it's still valuable for non-time-critical applications in domains such as office, home, hotel, school and other places which are not very crowded. Secondly, one method to improve performance is restrict context providing in a time interval. This will greatly decrease the reasoning burden and k–anonymity generating computation. But the side effect is the quality of context-awareness will be declined. Thirdly, centralized context server is a bottleneck for large smart space so a distributed mechanism is necessary.

Strongness and Weakness. Our middleware approach gets a global view of data and provides more possibility to protect data in a anonymous way. Because the computation ability restriction on handhold devices, policy here is strong to preserving privacy for data owner even if he does not even has a computable device. In addition, our policy support kinds of query, person based and attribute based, which will greatly satisfy different data users.

One weakness is *communication*. Communication between data owner and data collector, communication between data user and data collector will be vulnerable if a malicious third party wants to steal information. Using encrypted communication is one way. However, handhold devices with end users run at a lower CPU speed with very limited computation power. Asymmetric cryptography requires intensive computations. The RSA algorithm requires long integer calculations with numbers of 1000 to 2000 bit length. Therefore, the use of asymmetric cryptography should be kept to a minimum. If end device is powerful desktop, such secure communication can be applied.

Another is *User interfaces*. We have offered various types of policy rules to control person privacy policy, and this will make data owner enabled to handle kinds of situation. However, they also feel it is difficult to understand and customize their own policies. So we need a wizard like tool to direct and help policy instantiation.

Last but not least, *correlation between location and activity* is obvious. A context "Tom is having sports in campus" is blurred from "Tom is playing football in playground". With the first context, you may probably guess Tom is on playground because having sports are very related to playground. Although we can achieve $k-$anonymity on both location and activity, this correlation can not be eliminated easily.

7   Related Work

Policy based solutions are discussed broadly. Myles [4] established a middleware with distributed policy validators to control privacy. Hengartner [5] analyzes various of requirements of construct a access control to people's location information. Its characteristic is support of unknown users and transitivity control. Snekkenes [3] discusses the question of access control and presents a lattice model to define context accuracy which made sets of data relating to the same person is comparable. Marc Langheinrich gives principles and guidelines of designing privacy-aware applications and implement a privacy awareness system (pawS) based on P3P in [13] and [14] respectively. Gandon et al. [15] defines a sematic e-Wallet to supporting automated identification and access of personal resources and it also empower users to selectively control who has access to their contextual information and under which conditions. It has obfuscation rules for context, by abstraction(similar to our ontology based blurring) and falsification. The work of Hong et al. [16] analyze privacy requirements for end users and application developers and also presents policies for privacy control. The difference between our work and theirs are we considered more anonymity properties of context under a perspective of data collector and implemented a middleware based on a pervasive infrastructure both support policy access control and context accuracy generation.

For attack models, Timo Heiber and Pedro Marron [9] presents how to model inference-based attacks on context information. Alf Zugenmaier [17] provide a diamond like privacy model in mobile computing environment using relations between the users, devices, actions and locations. All such model can be the guidance for policy design.

For anonymous needs, Sweeney [10,18] proposes k-anonymity to provide privacy protection when sharing information on a large number of users with other parties in such a way that the users in the data set cannot be identified. Gruteser and Grunwald describe spatial and temporal cloaking to preserve $k-$anonymity [19]. They have a trusted middleware server, and people reported their location to it. The server gathers information and obfuscates location for users. However, their approach do not have a flexible identity controlling mechanism. Beresford and Stajano [8] describe mix zones for pervasive computing. Their experiments show that using anonymity set to measure privacy is not always effective, but we still believe it usable on the conditions that people do not move frequently or pervasive environment covers large area. And their work is only for location, but when it is applied to activity and other time relevant attributes, things may differ. More accurate metric, using entropy, gives us a new view of privacy processing on data. Hitchhiking [20] is a new approach that treats location as the primary entity of interest, because authors believe a reduction in precision can not be fit to all scenarios and hitchhiking works well that category of applications. LCP(Life-Cycle Policy) [21] is binded to datasets in order to regulating context data progressive degradation. In addition, it promise a one-way property of data which ensures that degraded information can no more be recovered from the current database content.

8   Conclusion and Future Work

In ubiquitous and pervasive computing, after data owner's information is collected, data collector should be responsible for protect the information from unintended disclosure, which we call data collector end protection for privacy compared to data owner end protection. Firstly, discussing challenges that arise when specifying privacy control policy in a pervasive environment. Secondly, design a contextual rule based access control policy mechanism enriched with methods of generating blurred context and guaranteeing information anonymous. Thirdly, we present implementation with an ontology based context model and evaluation of the middleware.

Future work includes on more effective mechanism to achieve anonymity with consideration of large volume of changed context. Other work should be concentrated on a more complex and adaptive disclosure according to history information.

Acknowledgement

The work is funded by 973 Project of China (2002CB312002) and NSFC (60403014), NSFJ(BK2006712).

References

1. Harper, R.H.R.: Why people do and don?t wear active badges: A case study. Computer Supported Cooperative Work **4**(4) (1995) 297–318
2. Li, J., Bu, Y., Chen, S., Tao, X., Lu, J.: Followme: On research of pluggable infrastructure for context-awareness. In: AINA (1), IEEE Computer Society (2006) 199–204
3. Snekkenes, E.: Concepts for personal location privacy policies. In: ACM Conference on Electronic Commerce, ACM (2001) 48–57
4. Myles, G., Friday, A., Davies, N.: Preserving Privacy in Environments with Location-Based Applications. IEEE Pervasive Computing **2**(1) (2003) 56–64
5. Hengartner, U., Steenkiste, P.: Protecting access to people location information. In Hutter, D., Müller, G., Stephan, W., Ullmann, M., eds.: SPC. Volume 2802 of Lecture Notes in Computer Science., Springer (2003) 25–38

6. Reiter, M.K., Rubin, A.D.: Crowds: Anonymity for web transactions. ACM Trans. Inf. Syst. Secur. **1**(1) (1998) 66–92
7. Goldschlag, D.M., Reed, M.G., Syverson, P.F.: Onion routing. Commun. ACM **42**(2) (1999) 39–41
8. Beresford, A.R., Stajano, F.: Location Privacy in Pervasive Computing. IEEE Pervasive Computing **2**(1) (2003) 46–55
9. Heiber, T., Marron, P.J.: Exploring the relationship between context and privacy. In Robinson, P., Vogt, H., Wagealla, W., eds.: Privacy, Security and Trust within the Context of Pervasive Computing. The Kluwer International Series in Engineering and Computer Science; 780, University of Stuttgart, Faculty of Computer Science, Electrical Engineering, and Information Technology, Springer-Verlag (2005) 0–0 ISBN 0-387-23461-6.
10. Sweene, L.: k-anonymity: A model for protecting privacy. International Journal of Uncertainty, Fuzziness and Knowledge-Based Systems **10**(5) (2002) 557–570
11. Wang, K., Yu, P.S., Chakraborty, S.: Bottom-up generalization: A data mining solution to privacy protection. In: ICDM, IEEE Computer Society (2004) 249–256
12. Sweeney, L.: Datafly: A system for providing anonymity in medical data. In Lin, T.Y., Qian, S., eds.: DBSec. Volume 113 of IFIP Conference Proceedings., Chapman & Hall (1997) 356–381
13. Langheinrich, M.: Privacy by design - principles of privacy-aware ubiquitous systems. In Abowd, G.D., Brumitt, B., Shafer, S.A., eds.: Ubicomp. Volume 2201 of Lecture Notes in Computer Science., Springer (2001) 273–291
14. Langheinrich, M.: A privacy awareness system for ubiquitous computing environments. In Borriello, G., Holmquist, L.E., eds.: Ubicomp. Volume 2498 of Lecture Notes in Computer Science., Springer (2002) 237–245
15. Gandon, F.L., Sadeh, N.M.: Semantic web technologies to reconcile privacy and context awareness. J. Web Sem. **1**(3) (2004) 241–260
16. Hong, J.I., Landay, J.A.: An architecture for privacy-sensitive ubiquitous computing. In: MobiSys, USENIX (2004)
17. Zugenmaier, A., Kreuzer, M., Müller, G.: The freiburg privacy diamond: An attacker model for a mobile computing environment. In Irmscher, K., Fähnrich, K.P., eds.: KiVS Kurzbeiträge, VDE Verlag (2003) 131–141
18. Sweene, L.: Achieving k-anonymity privacy protection using generalization and suppression. International Journal of Uncertainty, Fuzziness and Knowledge-Based Systems **10**(5) (2002) 571–588
19. Gruteser, M., Grunwald, D.: Anonymous usage of location-based services through spatial and temporal cloaking. In: MobiSys, USENIX (2003)
20. Tang, K.P., Keyani, P., Fogarty, J., Hong, J.I.: Putting people in their place: an anonymous and privacy-sensitive approach to collecting sensed data in location-based applications. In: CHI '06: Proceedings of the SIGCHI conference on Human Factors in computing systems, New York, NY, USA, ACM Press (2006) 93–102
21. Anciaux, N., van Heerde, H., Feng, L., Apers, P.: Implanting Life-Cycle Privacy Policies in a Context Database. Technical Report TR-CTIT-06-03, CTIT, University of Twente (2006)

Auditing and Inference Control for Privacy Preservation in Uncertain Environments

Xiangdong An[1,2], Dawn Jutla[2], and Nick Cercone[1]

[1] Faculty of Computer Science
Dalhousie University
Halifax, NS B3H 1W5, Canada
{xan, nick}@cs.dal.ca
[2] Department of Finance, Information Systems, and Management Science
Saint Mary's University
Halifax, NS B3H 3C3, Canada
{xan, dawn.jutla}@smu.ca

Abstract. In ubiquitous environments, context-aware agents have been developed to obtain, understand and share local contexts with each other so that the environments could be integrated seamlessly. Context sharing among agents should be made privacy-conscious. Privacy preferences are generally specified to regulate the exchange of the contexts, where who have rights to have what contexts are designated. However, the released contexts could be used to derive those unreleased. To date, there have been very few inference control mechanisms specifically tailored to context management in ubiquitous environments, especially when the environments are uncertain. In this paper, we present a Bayesian network-based inference control method to prevent privacy-sensitive contexts from being derived from those released in ubiquitous environments. We use Bayesian networks because the contexts of a user are generally uncertain, especially from somebody else's point of view.

1   Introduction

Mobile, embedded and ubiquitous communication and computing devices such as cell phones, PDAs, and laptops are broadly used in our lives. These devices use a wide range of heterogeneous networks and provide a wide range of services. Context-aware agents have been developed to integrate these facilities in a transparent way so that they could be available for different tasks at different times and in different locations [1,2,3]. The integrated environments are called ubiquitous (pervasive) environments [4]. Here, by context, we mean any information that can be used to characterize an entity in the environments such as locations, time, capabilities, lighting, noise levels, services offered and sought, activities and tasks engaged, roles, beliefs and the preferences [1,5]. Context-aware agents capture, interpret, reason about, and exchange such contextual information with each other for integrating the resources in the environments. However, not all contextual information can be exchanged unconditionally. The contexts of a user may contain or imply his privacy. Privacy is "the right to be left alone"[6], which can be divided into (1) bodily privacy, (2) territorial privacy, (3) communication privacy, and

P. Havinga et al. (Eds.): EUROSSC 2006, LNCS 4272, pp. 159–173, 2006.

(4) information privacy [7]. A user generally considers his contexts as his information privacy, which he may not like to share with others unconditionally. For example, John's current location, the car being driven, the time arriving at some business site, the goods bought and the time spent at the site could all be private or privacy containing. John may not want to disclose these contexts to others unconditionally. To control the disclosure of our privacy-sensitive contexts, we could specify privacy preferences to designate who have rights to have what contexts about us.

In the literature there exist some privacy preference specification platforms. For example, corresponding to P3P (the Platform for Privacy Preferences) [8], the preference language APPEL (A P3P Preference Exchange Language) [9] has been developed to help users express their preferences in information exchanging. An XPath-based [10] preference language, called XPref [11], has been proposed to improve APPEL (so that what is acceptable can be specified, etc). Nevertheless, these preference languages are not suitable for ubiquitous environments, where many different domains, systems, ontologies, and hence many different policies and preferences, need to exist together. For properly specifying and comparing various privacy preferences in ubiquitous environments, APPEL has been further extended to SWAPPEL (Semantic Web APPEL) [12,13,14]. Semantic e-Wallet [15] is another effort to improve APPEL using Semantic Web to ensure preferences of different contextual attributes can be specified over different ontologies. *Rei* is a language designed for preference specification in ubiquitous environments, which allows different entities to communicate about their preferences [16,17]. The language is not tied to any specific application and permits domain specific information to be added without modification [18].

Nevertheless, these preference languages do not provide a mechanism for inference control in preference specification. However, by inference, an adversarial agent could derive the privacy-sensitive contexts of the interested parties from those released. For example, John does not want anybody to know that he has a serious disease, but his agent tells Bob's agent that he is currently in ICU (intensive care unit) of a hospital. For another example, John may not want anybody to know his marital status but his agent could tell Bob's agent that he has a daughter but lives alone. A mechanism is needed to ensure the information to be released won't tell more than desired. Inference control [19,20,21] has been applied in statistical databases [22] to protect sensitive information from being derived, but none of them is particularly tailored for context protection in ubiquitous environments, in particular when environments are uncertain.

We say a context is *public* or *observable* if it can be observed by others easily (e.g. the weather of his city). We say a non-public context is *unclassified* if it is not privacy-sensitive by itself (e.g. one's preference on food), and *classified* if it is privacy-sensitive by itself (e.g. the diseases one has). We say an unclassified context is *releasable* or *disclosable* if it does not imply any classified contexts and *unreleasable* or *undisclosable* otherwise. The released contexts could be combined with the public ones (which could be none) to infer those not released, and even those classified. This work presents a method for auditing the unclassified contexts to be released in order to determine if the classified contexts could be inferred. In ubiquitous environments, contexts, either sensed or interpreted, could be ambiguous [23]. For instance, Bob very probably but does not necessarily have some diseases if he appears in a hospital. He could go there to see a friend or just for a blood test for a routine physical examination. For another instance, John may not exactly know how heavy the traffic on his road home will be to-day. Bayesian networks (BNs) [24] have been broadly used to effectively reason about the states of domains in many uncertain systems [25,26,27]. In this paper, we propose a Bayesian network-based reasoner to help a user audit his contexts to be released dynamically. The reasoner by inference ensures that all information that could be implied by the contexts to be disclosed is what expected. There exist a lot of other plausible reasoning formalisms that can deal with uncertainties such as default logic [28], defeasible reasoning [29], fuzzy logic [30], circumscription logic [31], autoepistemic logic [32], Dempster-Shafer calculus [33,34], and possibilistic logic [35]. However, we are bound to use probabilistic reasoning if we numerically represent uncertainty for a set of mutually exclusive and collectively exhaustive alternatives and want a coherent numerical model of human reasoning [36].

The paper is organized as follows. In Section 2, we review the related work. We give an introduction to Bayesian networks in Section 3. In Section 4, we first present a detailed discussion on contextual inference control in uncertain environments. We then give the problem a formal description. An algorithm is presented and shown to be secure. In Section 5, by an example, we show how Bayesian networks can be applied to the problem and their effectiveness in the application. Concluded remarks are made in Section 6.

2   Related Work

In (multilevel) statistical databases [37,38], inference control [39,40,21] has been used to protect information with a higher security level from being inferred. For exact inference control, many restriction based techniques have been studied which include restricting the size of query results [19,41], controlling the overlap of queries [42], suppressing sensitive data values [43], and auditing queries to determine whether inference is possible [44,45]. For statistical inference control, some perturbation based techniques have been proposed such as adding noises to source data [46] or to the query results [47], swapping values among records [48], and sampling data to answer queries [20]. However, these techniques are generally not applicable to context management in ubiquitous environments. This is because the set of contexts of an entity in a ubiquitous environment is usually much smaller in size and may only correspond to a record in databases, but inference control techniques for statistical databases need to manipulate multiple or huge records.

In particular, our work assume we answer a query by either presenting the correct answer or refusal. That is, we do not lie. The assumption is reasonable in some scenarios where the consequences of a false answer may result in expensive costs [49]. Also our work address the inference control problems in uncertain environments. We, as human beings, deal with lots of uncertainties in a daily basis and usually arrive at reasonable solutions, but to our knowledge, few of inference control techniques for uncertain domains have ever been reported. Our work presents an approach to audit the contexts to be released dynamically in order to determine if any inference is possible from them. In contrast to the static approach (e.g. [50]) where appropriate access rights are set up in advance (if possible) based on a complete analysis of the protection requirements, the dynamic approach (e.g. [51]) deals with each query based on the current query history.

It has been shown [52,49,53] in deterministic domains, for the most elementary queries, some strategies involving either lying or refusal or both are secure even when potential secrets or secrecies (both alternatives of a fact) are known. However, the proofs are made based on assumption that whether a true answer to a query implies some secrets is immediately known. That is, how to find if an answer implies some secrets is not discussed. Our paper presents a method to find if an answer implies some secrets in uncertain domains.

3   Bayesian Networks

Bayesian networks (BNs) have been widely accepted as an effective formalism for inference with uncertainty knowledge in artificial intelligent systems [54,55]. A BN is a directed acyclic graph (DAG) where each node represents a discrete domain variable of interest. Each node could take different values and is associated with a conditional probability distribution (CPD) specifying probabilities of the node taking specific values given the values of its parents. Variables without any parents are not influenced directly by other variables.

Definition 1. *A **Bayesian network** is a triplet $(V, G, \mathcal{P})$, where V is a set of variables, G is a connected DAG, $\mathcal{P}$ is a set of probability distributions: $\mathcal{P} = \{P(v|\pi(v)) \mid v \in V\}$, where $\pi(v)$ denotes parents of v in G.*

It can be shown that the joint probability distribution specified by $\mathcal{P}$ is:

$$P(V) = \prod_{v \in V} P(v|\pi(v)). \tag{1}$$

In a BN, the DAG structure encodes the causal dependencies among domain variables and the strength of the dependencies is quantified by the set of local conditional probability distributions (CPDs). The DAG structure makes the real dependencies in a problem domain explicit, where each node is conditionally independent of its non-descendants given its parents. This property allows us to solve complex problems by cheaper local computations.

Consider an example where a bird could either be healthy or be infected with bird flu. By a test, a bird doctor can determine with a high probability whether the bird is infected with bird flu or not. The situation can be represented by two random boolean variables, $F(lu)$ and $T(est)$. The variable F is *true* when the bird is infected with bird flu and *false* otherwise. The variable T is *true* when the outcome of the test is positive (claiming the bird is infected with bird flu) and *false* otherwise. Note that it is possible that the bird is healthy when the test has a positive outcome and vice versa. However, we assume the outcome of the test is highly dependent on the true state of the bird's health.

The example can be modeled by a Bayesian network as shown in Figure 1. The two random variables are represented by two nodes in the network. The arrow from F to

T indicates the causal relationship between the two respective variables. Each node is associated with a conditional probability distribution (the CPD associated with node F degenerates into a marginal probability distribution since F does not have any parents in the network). The CPD associated with T indicates the strong dependency between variables F and T. We assume we do not have any prior knowledge about if a bird is infected with bird flu or not. Hence, we assign a uniform distribution to variable F. In Bayesian probability theory, probabilities are subjective corresponding to the degree of belief of reasoners in the truth of the statements. The degree of belief is different from the degree of truth. People with different prior knowledge could *correctly* obtain different results from Bayesian probability theory.

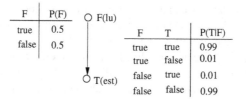

F	P(F)
true	0.5
false	0.5

F	T	P(T|F)
true	true	0.99
true	false	0.01
false	true	0.01
false	false	0.99

Fig. 1. A BN to determine whether a bird is infected with bird flu or not

In the BN as shown in Figure 1, variable T, which represents the outcome of the test, is often called *information* or *evidential* variable. The states of these variables are usually given or can be measured in a straightforward manner. Variable F, which represents the actual state of a bird's health, is called a *hypothesis* variable. The states of such variables cannot be obtained directly. However, BNs can, based on evidence collected from information variables, help efficiently calculate the probabilities of hypothesis variables taking different values. In our case, the doctor may want to determine the probability of a bird being infected with bird flu given the outcome of the test is positive. By calculation, the bird is infected with bird flu by 99% if the outcome of the test is positive. It is convincing given the belief that the two variables F and T are strongly dependent with each other (i.e. $P(T = true|F = true) = 0.99$ but $P(T = true|F = false) = 0.01$). Anyway, our belief does affect the reasoning results. For example, we may believe a bird collected from a non-epidemic area has a very low probability (e.g. 1%) to be infected with bird flu. Based on the belief, given the outcome of the test is positive, the probability of the bird being infected with bird flu will become 50%.

For contextual inference control, the information variables are associated with the public or released contexts, and the hypothesis variables are associated with the classified contexts (e.g. Bob's true activities at some location, etc). Our goal is to prevent the classified contexts from being derived from the observed values of information variables.

4   The Problem and the Method

In a ubiquitous environment, a user may have many contexts such as his location, the food he likes, the local weather, the restaurants he often visits, etc. Some of these contexts could be observed by the others (e.g. today's weather condition), some of them could be disclosed by the owner to the others (e.g. the makes of cars or painters he likes), and some of them may not be disclosed voluntarily by the owner (e.g. the professors he hates, some of the diseases he has or had and how often he visits hospitals). Nevertheless, based on the dependencies among one's contexts, the classified contexts could be inferred from those disclosed or public.

We may be able or need to reason about more contexts of a person based on our prior knowledge and new evidence about the user. Our prior knowledge about a user could come from common senses (e.g. a person should have had a natural father and a natural mother, a user will attend classes from time to time if he is a student) or from built up information about the person (e.g. he is generally honest, he attends a class 3:30pm-4:30pm on Wednesdays). New evidence could be from our observation about the user (e.g. today's weather of the user's city is good, the user is attending a conference, the user looks downhearted, etc), or from the information the user has disclosed to us (e.g. the user will attend a class tonight, or attended a class last night). As an example, suppose Alice does not like her agent to release any of her course marks to her friends, but she does allow her agent to disclose what courses she is going to take in the next term so that potential pleasant classmates could be found as early as possible. In this case, our prior knowledge could be: a student cannot take the same course any more once he has passed it. Our evidence could be: Alice attended the classes of course A in the last term; Alice plans to take course A in the next term. Though other agents may not know Alice's exact mark on the course A in last term, by inference they could know Alice may have failed the course.

If our prior knowledge on all contexts about a person is accurate and certain, we would be able to reason about the state of the domain deterministically. For the example above, if that Alice attended the classes for the course A indicates she had registered and had not quit the course in the last term, Alice should have failed the course. However, this is not always true in the real world. There generally always exist some exceptions to a rule or a statement. In this example, that Alice attended the classes for the course A does not necessarily mean she had registered the course or she had not quit it. For another example, one could still be considered healthy if he is becoming thin due to his being on diet or his exercising. We may need to figure out the credibility of the conclusions derived from our uncertain knowledge. In uncertain domains, without help of a rigorous model, an agent may have difficulty to determine if a classified context has been given away by those disclosed. Bayesian networks (BNs), based on Bayesian probability theory, provide a framework that can deal with uncertainties coherently and efficiently. In this paper, we propose to apply BNs to dynamic inference control in context management in uncertain ubiquitous environments.

As mentioned above, we assume a user may have some contexts that is observable to the others (e.g. the weather condition), called *observable* or *public* contexts. The remaining contexts that are unobservable to the others could be divided into the *unclassified* contexts, which are privacy-insensitive by themselves, and the *classified* contexts, which are privacy-sensitive by themselves. In general, the problem of contextual inference control can be summarized formally as follows in Problem statement 2. For a set S, we use $|S|$ to represent the cardinality of S.

Problem Statement 2. *Assume a user has 3 sets of contexts of different natures: the set of observable contexts* $S_o = \{c_{o0}, c_{o1}, ..., c_{oi}\}$ ($|S_o| \geq 0$), *the set of unclassified contexts* $S_u = \{c_{u0}, c_{u1}, ..., c_{ui}\}$ ($|S_u| \geq 0$), *and the set of classified contexts* $S_c = \{c_{c0}, c_{c1}, ..., c_{ci}\}$ ($|S_c| \geq 1$). *We assume no significant information about* S_c *could be inferred from* S_o *based on prior knowledge* K, *but all or part of* S_c *could be derived based on* S_o, *a set* R *of released unclassified contexts* ($R \subseteq S_u$), *and* K. *Our problem is to determine if a subset* R *of* S_u ($R \subseteq S_u$) *could be released without endangering the security of the contexts in* S_c *(i.e. no contexts in* S_c *could be derived from the combination of* S_o, R *and* K).

Example 3 below illustrates the problem.

Example 3. *Alice does not want anybody to know she is sick. We represent the classified context (Alice is sick) by* c_j ($\in S_c$). *Before any contextual evidence is obtained, the probability of Alice being sick could be very low based on prior knowledge* K: $P(c_j|K) = 5\%$. *Nevertheless, with the set of observable contexts about Alice considered (e.g. she looked tired yesterday), the probability of Alice being sick could become* $P(c_j|S_o, K) = 30\%$. *After we get the released context information* R ($R \subseteq S_u$) *from her (e.g. she is in a medical center), our belief about* c_j *could become* $P(c_j|S_o, R, K) = 80\%$. *Our objective is to figure out if a subset* R *of* S_u ($R \subseteq S_u$) *would, based on* K, *significantly imply* c_j *when released and considered with* S_o.

Uncertain reasoning is generally non-monotonic. That is, new evidence could lead to knowledge that contradicts the old. Depending on the contents of S_o and S_u, it is possible that before one more evidence $e \in S_u$ is obtained and considered, our belief is $P(c_j|S_o, R, K) = 80\%$ ($e \notin R$), but after e is considered, $P(c_j|S_o, R \cup \{e\}, K) = 20\%$ (e.g. for the instance above, if Alice looked tired yesterday $\in S_o$, she did not come to work today $\in R$, and e="she starts her vacation today"). This is because of "explaining away" [24] in uncertain reasoning: when new evidence is found to be able to explain the same hypothesis, the old explanation would become less credible. That Alice is sick could explain that she did not come to work today, but that she starts her vacation today would make the previous explanation less credible. That is, it is possible that the contexts in S_c that are implied by a subset R of contexts in S_u are denied when more contexts in S_u are released and considered, and vice versa. Therefore, when figuring out whether a set of unclassified contexts could imply the classified ones, the set of unclassified contexts should be considered as a package regarding their effects in disclosing classified contexts.

To calculate if a subset of unclassified contexts implies classified ones, we may need a confidence threshold t, say $t = 85\%$. If the reasoning indicates that some classified context may exist by a probability over t based on the unclassified contexts to be released (i.e. significant information about S_c is implied), the corresponding set of unclassified contexts is considered *unreleasable* and cannot be released together.

In deterministic domains, a context can be represented by a deterministic statement that is true such as Mr. Smith has AIDS, Mr. Johnson is not on campus, etc. However, in uncertain domains we may have difficulty to judge if a statement is absolutely true or not. To reflect the uncertainty in our knowledge and for formal knowledge representation, without losing generality, we may use a boolean random variable to represent a deterministic statement which is either *true* or *false*, and assign a probability to the truth of the statement based on our belief. For instance, Mr. Smith has AIDS by 70%. We say a context is an *instantiation* of a context random variable and a context random variable represents a context *variety*. The space D_c of a context variable (variety) c is $D_c = \{\text{true, false}\}$. For a set of contexts S, its space D_S is the Cartesian product of the spaces of variables in S, namely $D_S = \prod_{s \in S} D_s$. Each element in D_S is called a *configuration* of S, which corresponds to a set of contexts. For a set of context varieties, only one of its configurations actually holds at one time, but in uncertain domains, every configuration would be assigned a probability. The configuration of a set S of contexts that actually holds is denoted by f_S. With more implementation details considered, the problem described in Problem statement 2 could be specified more exactly as in Problem statement 4.

Problem Statement 4. *In a ubiquitous uncertain environment E, each context-aware agent A_i corresponds to a user U_i in E $(0 < i \leq n, n \geq 2)$. A user U_i has 3 sets of context varieties of different natures: the set of observable context varieties C_i^o $(|C_i^o| \geq 0)$, the set of unclassified context varieties C_i^u $(|C_i^u| \geq 0)$, and the set of classified context varieties C_i^c $(|C_i^c| \geq 1)$. A_i has access to the most exact knowledge [1] K_i about C_i^o, C_i^u, C_i^c and their relationships, but only has access to some knowledge K_{ij} [2] about C_j^o, C_j^u, C_j^c, and their relationships $(0 < j \leq n, j \neq i)$. All or part of $f_{C_i^c}$ could be derived by an agent A_k based on $f_{C_i^o}$, $f_{R_{ki}}$ $(R_{ki} \subseteq C_i^u)$ that has been released to A_k, and K_{ki} $(0 < k \leq n, k \neq i)$. Before releasing f_R $(R \subseteq C_i^u \setminus R_{ki})$, agent A_i, based on its guess G_{ik} about agent $A_k's$ knowledge K_{ki}, needs to figure out:*

1. *If any element of $f_{C_i^c}$ could have been inferred [3] by A_k from its knowledge about $f_{C_i^o}$ and the previously released $f_{R_{ki}}$ (could be $\emptyset$) based on K_{ki};*
2. *Whether f_R could be released to A_k without endangering the security of $f_{C_i^c}$ (i.e. no any element of $f_{C_i^c}$ could be inferred by A_k).*

To preserve secrets, our strategy has to be based on the assumed knowledge of the adversarial agents [52]. In Problem statement 4, the assumed knowledge of A_k about U_i is G_{ik}. The closer G_{ik} is to K_{ki}, the closer the state of domain derived based on G_{ik} is to that understood by A_k. In general, the assumed knowledge of an adversary A_k about a user U_i $(k \neq i)$ would include all public contexts of U_i and any insensitive contexts of U_i that have previously been released to A_k. If it is possible that a group Q of adversarial agents may share their information about user U_i, the assumed knowledge of an agent $A_q \in Q$ about U_i could include all insensitive contexts of U_i that have been disclosed to the agents in the group. Anyway, the assumption could be made either conservatively or imprudently. The most conservative assumption is that $G_{ik} = K_i = K_{ki}$. The most imprudently assumption could be limited to some obvious knowledge about contexts of U_i. Hence, A_i could pre-evaluate the upper bound and the lower bound of the state of the domain to be understood by A_k before releasing any contexts in C_i^u to A_k.

---

[1] Exact knowledge could still be uncertain.

[2] K_{ij} is generally less exact than K_j.

[3] In an uncertain environment, a statement is *inferred* if the statement is shown to be true by a probability over a confidence threshold.

With a BN based context model, we can examine whether a set of unclassified contexts, once released and considered together with the public and previously released contexts if any, could imply the classified contexts. A BN model corresponding to a user can be constructed based on the assumed knowledge about the user, or learned automatically from the available data about the user. There are a lot of work about BN learning in literatures [56,57,58,59,60]. By BN learning and inference, a BN model can be obtained and improved over time.

A BN-based algorithm for dynamic inference control is proposed as Algorithm 5, where agent $A'_i s$ guess G_{ik} about prior knowledge K_{ki} of A_k on U_i is modeled as a BN. Note in Algorithm 5, we refuse to disclose f_R not only when f_R implies any classified contexts, but also when any negation of f_R implies any classified contexts. This is because in our case we answer a query by either giving a correct answer or refusal, and when we refuse, an informed and sophisticated user may wonder why [52]. If we refuse only when f_R implies any classified contexts, the user could check what sensitive contexts f_R would imply and hence find them. However, if the user knows our strategy is refusal whenever any instantiation of R could imply private contexts, he won't benefit from our refusal since he needs to check what sensitive contexts could be indicated by all configurations of R, which he can do before being refused. We further explain it by Example 6.

Algorithm 5. *A BN model M is constructed based on G_{ik}. Before A_i sends A_k f_R of a subset R of unclassified contexts C_i^u about U_i, by the following operations, A_i figures out if f_R will be able to reveal any contexts in $f_{C_i^c}$ to A_k when considered together with $f_{C_i^o}$ and $f_{R_{ki}}$. Let P_m be the probability of context m being true. Let T be the probability confidence threshold.*

1 enter evidence $f_{C_i^o}$, $f_{R_{ki}}$ to M;
2 for each possible configuration of R
3   enter the configuration to M;
4   for each context $m \in f_{C_i^c}$
5   if $P_m \geq T$, f_R cannot be released and return;
6 f_R can be released and return;

Example 6. *Suppose "Mr. Smith has AIDS" is a classified context. To protect the secret, when the answer is positive, we would refuse to answer a question such as "Has Mr. Smith visited a doctor for AIDS?". However, if our strategy is refusal only when the answer is "Yes", the enquirer will know the truth once he gets refused. Anyway, if our strategy is refusal on either "Yes" or "No", the enquirer gets nothing from the query. Hence, in Algorithm 5, not only f_R, but also the other configurations of R need be tested for the security of the classified contexts.*

To find if f_R is releasable or not needs to check the probability of each context in $f_{C_i^c}$ and all configurations of R. Hence, the computational complexity is linear to $|C_i^c|$ but exponential to $|R|$ (i.e. $O(2^{|R|}|C_i^c|)$). For an elementary query which involves only one context, the computation would be linear to $|C_i^c|$ (i.e. $O(|C_i^c|)$). The computation would be effective when the number of contexts involved in a query is not large. It is fortunate that a query generally does not involve too many contexts. Note the two questions in

Problem statement 4 are very similar when they are treated separately: The first question examines if any secrets have been detected, and the second question examines if any secrets will be detected. However, due to the non-monotonicity of uncertain reasoning, when the two questions are considered together, we get chances to cover up the secrets that could have been implied.

From analysis above, we get Proposition 7.

Proposition 7. *Algorithm 5 will halt and the released unclassified contexts won't imply any classified contexts in the sense of indicating any classified contexts to be true by a probability over a specified confidence threshold.*

As mentioned above, BNs have been broadly proved to be effective in inference in uncertain domains [24,54,61,55,27,25]. In the next section, we demonstrate their effectiveness in the application by an example.

5   Demonstration

In this section, by an example, we show how Bayesian networks can be applied to protecting classified contexts from being inferred and their effectiveness.

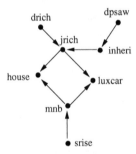

Fig. 2. A BN for reasoning about the privacy-sensitive contexts

Suppose John works and does an excellent job in a company. Everybody guesses his salary would be raised if possible (25%). Assume John's father has severe health problems, and could die by a probability of 5%. John will very probably (95%) get inheritance from his father if his father passes away. Recently John not only got a brand new car but also bought a new house. Based on the observation, his colleagues think John's salary should have been raised, and raised greatly. Nevertheless, John's colleagues realize that this could be because that John has got a large inheritance from his father when John told them his dad has died, though they are not very sure (say, 50%) if John's father was rich. We assume John will have much more monthly surplus (balance) if his salary is raised. John will probably buy a new house and a good car if he has more money available.

We model this example using a Bayesian network as shown in Figure 2, where "drich" denotes "John's dad is rich" (true or false), "jrich" "John becomes rich" (true or false), "inheri" "John gets inheritance from his father", "dpsaw" "John's dad passes away" (true or false), "house" "John has a new house" (true or false), "luxcar" "John has a new car" (true or false), "mnb" "John has more monthly balance" (true or false), and "srise" "John's salary is raised" (true or false). The corresponding conditional probability distributions are specified, based on his colleagues' knowledge about him, in Tables 1, 2, and 3 respectively. The dot '.' in Table 3 denotes the corresponding conditions abbreviated (e.g. $P(jrich|.) = P(jrich|inheri, drich)$). John's colleagues can use this model to reason about what have happened to John, and John's agent can use the model to compute how John's colleagues would think about John.

Table 1. Marginal probability distributions of the variables without any parents

.	$P(drich)$	$P(dpsaw)$	$P(srise)$
true	0.5	0.05	0.25
false	0.5	0.95	0.75

Table 2. Conditional probability distributions of *inheri* and *srise*

dpsaw	inheri	$P(inheri|dpsaw)$	srise	mnb	$P(mnb|srise)$
true	true	0.95	true	true	0.9
true	false	0.05	true	false	0.1
false	true	0.0	false	true	0.01
false	false	1.0	false	false	0.99

Table 3. Conditional probability distributions of *jrich*, *house*, and *luxcar*: $P(jrich|inheri, drich)$, $P(house|jrich, mnb)$, and $P(luxcar|jrich, mnb)$

inheri	drich	jrich	$P(jrich|.)$	jrich	mnb	house	$P(house|.)$	jrich	mnb	luxcar	$P(luxcar|.)$
true	true	true	0.9	true	true	true	0.9	true	true	true	0.95
true	true	false	0.1	true	true	false	0.1	true	true	false	0.05
true	false	true	0.1	true	false	true	0.7	true	false	true	0.75
true	false	false	0.9	true	false	false	0.3	true	false	false	0.25
false	true	true	0.1	false	true	true	0.75	false	true	true	0.8
false	true	false	0.9	false	true	false	0.25	false	true	false	0.2
false	false	true	0.1	false	false	true	0.0	false	false	true	0.0
false	false	false	0.9	false	false	false	1.0	false	false	false	1.0

Suppose whether John drives a good car and lives in a new house are his observable context varieties (which his colleagues can observe easily). His unclassified context varieties are if his dad is alive, if his dad is rich, if he has inherited a large fortune from his father, and if he has become rich because of the inheritance. The classified context varieties are if his salary has been raised, and if he has much more monthly surplus due to his high salary.

Before any evidence is observed, the model indicates that John's dad has a probability of 50% to be a rich man; John has a low chance (4.75%) to inherit a fortune from his father because his dad has a low probability (5%) to die; John has a probability of 11.9% to become rich via inheritance or any other non-salary incomes (e.g. lottery winnings); John's salary could be greatly raised by a chance of 25%, and John has a chance of 23.25% to have much more monthly surplus on account of the chance of the increase of his salary; the chances that John could buy a good house and a good car are 24.25% and 25.86% respectively (John could buy a new house and a good car only when he comes into a large fortune or has much more monthly surplus from his monthly salary). These probabilities are consistent with the knowledge of John's colleagues about him.

Anyway, after his colleagues observe his new car and new house, they highly (73.72%) believe his salary has been increased greatly (their belief on John's having high monthly balance consequently rises to 75.35%), though their belief on John's coming into a large fortune also rises to 36.81% (that on John's having inherited a fortune from his father becomes 9.05%). Nevertheless, when John tells his colleagues that his dad has died, their belief on John's coming into a large fortune becomes 79.93% (that on John's inheritance of a fortune rises to 97.43%), that on John's salary increase becomes 46.72% (quite unconfident), and that on John's having more monthly surplus due to his high salary reduces to 46.48%. However, if John also releases the information that his dad was not rich, his colleagues' belief on his salary increase would become 76.49%, that on his having more monthly surplus rises to 78.30%, that on his becoming rich through inheritance or lottery winnings, etc becomes 32.40%, and that on his having inherited a fortune from his father is still high (95%) since his dad has died.

Hence, the unclassified contexts (John's father was not rich in this case), once released and combined with the observable contexts (John has a new house and a good new car in this case) and other released unclassified contexts (John's dad has died), could endanger the security of the classified contexts (John's salary has been increased and John has more monthly surplus). However, when only the unclassified context that John's father has died is released, John's classified contexts can even be covered up.

Therefore, due to the non-monotonicity of the uncertain reasoning, each new context released could enhance but could also weaken or retract the previous beliefs. To ensure no classified contexts could be derived from those released, the unclassified contexts should be examined dynamically and be released in a package. The impacts of the public and the previously disclosed unclassified contexts need be considered. A set of unclassified contexts can be released only when they, combined with the public contexts and the previously released privacy-insensitive contexts, won't disclose any (new) classified contexts.

6   Conclusion

In this paper, we propose a dynamic inference control approach to protect the sensitive contexts from being derived in uncertain environments. The problem of inference control in uncertain ubiquitous environments is formally described first. A Bayesian network-based algorithm is then presented to solve the problem. The approach is shown secure in protecting classified contexts in the sense of not implying any classified contexts to be true by a probability over a threshold. An example is presented to demonstrate how BNs can be applied to the problem and their effectiveness in the application.

References

1. Chen, H., Finin, T., Joshi, A.: An ontology for context-aware pervasive computing environments. Knowledge Engineering Review, Special Issue on Ontologies for Distributed Systems **18**(3) (2004) 197–207

2. Khedr, M., Karmouch, A.: Negotiating context information in context-aware systems. IEEE Intelligent Systems **19**(6) (2004) 21–29

3. Khedr, M., Karmouch, A.: ACAI: Agent-based context-aware infrastructure for spontaneous applications. Journal of Network and Computer Applications **28**(1) (2005) 19–44

4. Davies, N., Gellersen, H.W.: Beyond prototypes: Challenges in deploying ubiquitous systems. IEEE Pervasive Computing **1**(1) (2002) 26–35

5. Dey, A.: Understanding and using context. Personal and Ubiquitous Computing **5**(1) (2001) 4–7

6. Warren, S., Brandeis, L.: The right to privacy. Harvard Law Review **4** (1890) 193–220

7. An, X., Jutla, D.: A survey of privacy technologies. Technical report, Faculty of Computer Science, Dalhousie University, Halifax, NS, Canada (2005)

8. Cranor, L., Langheinrich, M., Marchiori, M., Presler-Marshall, M., Reagle, J.: The platform for privacy preferences 1.0 (P3P 1.0) specification. Technical report, W3C Recommendation, http://www.w3.org/TR/P3P (2002)

9. Cranor, L., Langheinrich, M., Marchiori, M.: A P3P preference exchange language 1.0 (APPEL 1.0). Technical report, W3C Working Draft, http://www.w3.org/TR/P3P-preference (2002)

10. Clark, J., DeRose, S.: XML Path language (XPath) Version 1.0. Technical report, W3C Recommendation, http://www.w3.org/TR/xpath (1999)

11. Agrawal, R., Kieman, J., Srikant, R., Xu, Y.: An XPath-based preference language for P3P. In: Proceedings of the 12th International WWW Conference (WWW'03), Budapest, Hungary (2003)

12. McBride, B., Wenning, R., Cranor, L.: A RDF schema for P3P. Technical report, W3C Note, http://www.w3.org/TR/p3p-rdfschema (2002)

13. Hogben, G.: P3P using the semantic web (OWL ontology, RDF policy and RDQL rules). Technical report, W3C Working Group Note, http://www.w3.org/P3P/2004/040920_p3p-sw.html (2004)

14. Hogben, G.: Describing the P3P base data schema using OWL. In Kagal, L., Finin, T., Hendler, J., eds.: Policy Management for the Web (PM4W): Proceedings of the 14th World Wide Web Conference (WWW'05) Workshop, Chiba, Japan (2005) 44–51

15. Gandon, F.L., Sadeh, N.M.: Semantic web technologies to reconcile privacy and context awareness. Journal of Web Semantics **1**(3) (2005)

16. Kagal, L., Finin, T., Joshi, A.: A policy language for pervasive systems. In: Proceedings of the 4th IEEE International Workshop on Policies for Distributed Systems and Networks (POLICY'03), Lake Como (2003)

17. Kagal, L., Finin, T., Joshi, A.: Declarative policies for describing web services capabilities and constraints. In: Proceedings of the W3C Workshop on Constraints and Capabilities for Web Services, Redwood Shores, CA, USA (2004)

18. Kolari, P., Ding, L., Ganjugunte, S., Kagal, L., Joshi, A., Finin, T.: Enhancing web privacy protection through declarative policies. In Sahai, A., Winsborough, W.H., eds.: Proceedings of the 6th IEEE International Workshop on Policies for Distributed Systems and Networks (POLICY'05), Stockholm, Sweden, IEEE Computer Society (2005) 57–66

19. Fellegi, I.: On the question of statistical confidentiality. Journal of American Statistical Association **67**(337) (1972) 7–18
20. Denning, D.: Secure statistical databases with random sample queries. ACM Transactions on Database Systems **5**(3) (1980) 291–315
21. Staddon, J.: Dynamic inference control. In Zaki, M.J., Aggarwal, C.C., eds.: Proceedings of the 8th ACM SIGMOD Workshop on Research Issues in Data Mining and Knowledge Discovery (DMKD'03), San Diego, CA, ACM Press (2003) 94–100
22. Shoshani, A.: Statistical databases: Characteristics, problems and some solutions. In: Proceedings of the 8th International Conference on Very Large Databases (VLDB'82), Mexico City, Mexico (1982) 208–213
23. Dey, A., Mankoff, J., Abowd, G., Carter, S.: Distributed mediation of ambiguous context in aware environments. In Beaudouin-Lafon, M., ed.: Proceedings of the 15th Annual ACM Symposium on User Interface Software and Technology (UIST'02), Paris, France, ACM Press (2002) 121–130
24. Pearl, J.: Probabilistic Reasoning in Intelligent Systems: Networks of Plausible Inference. Morgan Kaufmann Publishers, San Franciso, CA, USA (1988)
25. Wong, W.K., Cooper, G., Wagner, M.: Bayesian network anomaly pattern detection for disease outbreaks. In: Proceedings of the 20th International Conference on Machine Learning (ICML-2003), Washington DC, USA (2003)
26. Johansen, K., Lee, S.: Network security: Bayesian network intrusion detection. Technical report, Department of Computer Science, Johns Hopkins University, Baltimore, MD, USA (2003)
27. Kruegel, C., Mutz, D., Robertson, W., Valeur, F.: Bayesian event classification for intrusion detection. In: Proceedings of the 19th Annual Computer Security Applications Conference, LasVegas, Nevada, USA (2003)
28. Reiter, R.: A logic for default reasoning. Artificial Intelligence **13** (1980) 81–132
29. Nute, D.: Defeasible reasoning. In: Proceedings of the 20th Hawaii International Conference on System Science, Kailua-Kona, HI, USA, IEEE Press (1987) 470–477
30. Zadeh, L.: Fuzzy sets. Information and Control **8** (1965) 338–353
31. McCarthy, J.: Circumscription — a form of non-monotonic reasoning. Artificial Intelligence **13** (1980) 27–39
32. Moore, R.C.: Semantical considerations on non-monotonic logic. Artificial Intelligence **28** (1985) 75–94
33. Dempster, A.P.: Upper and lower probabilities induced by a multivalued mapping. Annual Mathematical Statistics **38** (1967) 325–339
34. Shafer, G.: A Mathematical Theory of Evidence. Princeton University Press, Princeton, MA, USA (1976)
35. Dubois, D., Lang, J., Prade, H.: Automated reasoning using possibilistic logic: semantics, belief revision, and variable certainty weights. IEEE Transactions on Knowledge and Data Engineering **6**(1) (1994) 64–71
36. Neapolitan, R.E.: Probabilistic Reasoning in Expert Systems: Theory and Algorithms. John Wiley & Sons, Inc., New York, NY, USA (1990)
37. Jajodia, S., Sandhu, R.: Polyinstantiation integrity in multilevel relations. In: Proceedings of the 1990 IEEE Symposium on Security and Privacy, Oakland, CA, IEEE Computer Society (1990) 104–115
38. Cuppens, F., Gabillon, A.: Logical foundations of multilevel databases. Data & Knowledge Engineering **29**(3) (1999) 199–222
39. Denning, D.E., Schlörer, J.: Inference control for statistical databases. IEEE Computer **16**(7) (1983) 69–82
40. Yip, R.W., Levitt, K.N.: Data level inference detection in database systems. In: Proceedings of the 11th IEEE Computer Security Foundations, Rockport, MA (1998) 179–189

41. Denning, D.E., Denning, P.J., Schwartz, M.D.: The tracker: a threat to statistical database security. ACM Transactions on Database Systems **4**(1) (1979) 76–96
42. Dobkin, D., Jones, A., Lipton, R.: Secure databases: Protection against user influence. ACM Transactions on Database Systems **4**(1) (1979) 97–106
43. Cox, L.H.: Suppression methodology and statistical disclosure control. Journal of the American Statistical Association **75**(370) (1980) 377–385
44. Chin, F.Y., Özsoyoglu, G.: Auditing and inference control in statistical databases. IEEE Transactions on Software Engineering **8**(6) (1982) 574–582
45. Kleinberg, J., Papadimitriou, C., Raghavan, P.: Auditing boolean attributes. In: Proceedings of the 19th ACM SIGMOD-SIGART Symposium on Principles of Database Systems (PODS'00), Dallas, TX, ACM Press (2000) 86–91
46. Traub, J.F., Yemini, Y., Woznaikowski, H.: The statistical security of a statistical database. ACM Transactions on Database Systems **9**(4) (1984) 672–679
47. Beck, L.L.: A security mechanism for statistical databases. ACM Transactions on Database Systems **5**(3) (1980) 316–338
48. Reiss, S.P.: Practical data-swapping: The first steps. ACM Transactions on Database Systems **9**(1) (1984) 20–37
49. Biskup, J., Bonatti, P.A.: Lying versus refusal for known potential secrets. Data & Knowledge Engineering **38** (2001) 199–222
50. Sicherman, G.L., de Jonge, W., van de Riet, R.P.: Answering queries without revealing secrets. ACM Transactions on Database Systems **8**(1) (1983) 41–59
51. Stickel, M.E.: Elimination of inference channels by optimal upgrading. In: Proceedings of the 1994 IEEE Symposium on Security and Privacy, Oakland, CA, IEEE Computer Society (1994) 168–174
52. Biskup, J.: For unknown secrecies refusal is better than lying. Data & Knowledge Engineering **33** (2000) 1–24
53. Biskup, J., Bonatti, P.: Controlled query evaluation for known policies by combing lying and refusal. Annals of Mathematics and Artificial Intelligence **40**(1-2) (2004) 37–62
54. Jensen, F.V.: An introduction to Bayesian networks. UCL Press, London, UK (1996)
55. Castillo, E., Gutierrez, J.M., Hadi, A.S.: Expert Systems and Probabilistic Network Models. Springer (1997)
56. Heckerman, D.: A tutorial on learning with Bayesian networks. Technical report, Microsoft Research, MSR-TR-95-06 (1995)
57. Tong, S., Koller, D.: Active learning for parameter estimation in Bayesian networks. In Leen, T.K., Dietterich, T.G., Tresp, V., eds.: Advances in Neural Information Processing Systems 13 (NIPS'00), Denver, CO, USA, MIT Press (2000)
58. Cheng, J., Greiner, R.: Learning Bayesian belief network classifiers: algorithms and systems. In: Proceedings of the 14th Canadian Conference on Artificial Intelligence, Ottawa, ON, Canada, Springer (2001) 141–151
59. Neapolitan, R.E.: Learning Bayesian Networks. Prentice Hall (2003)
60. Meng, D., Sivakumar, K., Kargupta, H.: Privacy sensitive Bayesian network parameter learning. In: Proceedings of the 4th IEEE International Conference on Data Mining (ICDM'04), Brighton, UK, IEEE Computer Society Press (2004) 427–430
61. D'Ambrosio, B.: Inference in Bayesian networks. AI Magazine **20**(2) (1999) 21–36

Developing a Context-Aware System for Providing Intelligent Robot Services*

Chung-Seong Hong[1], Joonmyun Cho[1], Kang-Woo Lee[1], Young-Ho Suh[1], Hyun Kim[1], and Hyun-Chan Lee[2]

[1] Electronics and Telecommunication Research Institute
Intelligent Robot Research Division, 161 Gajeongdong, Yuseonggu
Daejeon, Republic of Korea
{cshong, jmcho, kwlee, yhsuh, hyunkim}@etri.re.kr
[2] Hongik Univ., Dept. of Information and Industrial Eng., Sangsudong, Mapogu
Seoul, Republic of Korea
hclee@hongik.ac.kr

Abstract. To realize the vision of ubiquitous computing, it is important to develop a context-aware system which can help ubiquitous agents, services, and devices become aware of their contexts. Context-aware system is required to be capable of configuring appropriate context model and managing it through the context manipulation. In this paper, we propose the context space and the conceptual structure for context acquisition, representation, and utilization. We also describe how this conceptual structure is implemented in the service robot area. As the result of the research, we developed the CAMUS (Context-Aware Middleware for URC Systems) which is a context-aware server framework for a network-based intelligent robot system and applied it to the field test.

1 Introduction

Ubiquitous computing is an emerging paradigm of personal computing, characterized by the shift from dedicated computing machinery to ubiquitous computing capabilities embedded in our everyday environments. In the vision of ubiquitous computing environment, computer system understands their situational contexts and provides appropriate services to users [1]. To realize this vision, it is important to develop a context-aware system which can help ubiquitous agents, services, and devices become aware of their contexts because such computational entities need to adapt themselves to changing situations [2, 3].

Context is an available, detectable, and relevant piece of information or attribute about the circumstances, objects, or conditions surrounding a user at the time of interaction between the user and the computing environment. In contrast to the traditional computing paradigm, in which the computing environment is often static and well-defined, in Ubiquitous Computing the underlying environment is open and dynamic [4]. In an open and dynamic environment, contexts in question are often distributed in heterogeneous sources.

---

* This work was supported in part by MIC & IITA through IT Leading R&D Support Project.

P. Havinga et al. (Eds.): EUROSSC 2006, LNCS 4272, pp. 174–189, 2006.

As context-aware computing gets more and more interests, attempts applying it to diverse kinds of research fields, such as service robots, have been tried. In order that service robots can be introduced in our daily lives as if they were digital appliances and information devices, they should be able to provide various services that a user wants with context information. However, the functionalities of current service robots are neither worth their costs, nor do they have killer applications.

Under this background, ETRI (Electronics and Telecommunications Research Institute) has developed a new concept of a network-based intelligent robot, which is called as URC (Ubiquitous Robotic Companion) [5]. A network-based robot proposed in the URC can sufficiently utilize robot's own functionalities such as sensing, processing and acting by distributing its internal abilities to the external resources through the network as shown in Fig. 1. That is, it is enabled to utilize external sensors embedded in the environment to expand internal sensors of the robot itself. Also, a number of robots can use the high performance remote servers to increases the internal computing power. The improvement of sensing abilities brings enhancement to context awareness about external environments and users.

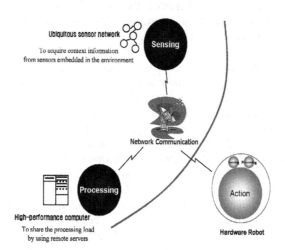

Fig. 1. The concepts of the URC

In order to realize this concept of a network-based service robot system, it is essential for a robot which is minimally equipped with its internal abilities to give users intelligent services by utilizing central servers. That is, when a user requests a service, it should be able to provide appropriate services by recognizing current contexts which are given by the central servers.

Thus, we have to understand about how to use the general context concepts in the development of context-aware system with the URC concept. Because most of context-aware applications are needed to use diverse kinds of context information, context-aware system is required to be capable of collecting, representing and utilizing appropriate context information.

In this paper, we propose the context space and the layered conceptual architecture for context acquisition, representation, and utilization. We also describe how this conceptual structure was implemented as a context-aware system in the service robot area. As the result of the research, we developed the CAMUS (Context-Aware Middleware for URC Systems), that is a context-aware server framework for a network-based intelligent robot system, and applied it to the field test.

The remainder of the paper is organized as follows. Section 2 discusses related researches and our contributions. In section 3, we describe the layered conceptual context structure and in section 4, we apply it to developing a context-aware system. Finally, in section 5, we conclude this paper with some remarks.

2 Related Work

Context-aware systems can be implemented in many ways. The approach depends on special requirements and conditions such as the location of sensors, the amount of possible users, the available resources of the used devices or the facility of a further extension of the system [6]. Furthermore, the method of context-data acquisition is very important when designing context-aware systems because it predefines the architectural style of the system at least to some extent.

Chen presents three different approaches on how to acquire contextual information [7]. The first approach is to use direct sensor access. This approach is used in devices with sensors locally built in. The client software gathers the desired information directly from these sensors. The second approach is to use a middleware infrastructure. The middleware based approach introduces a layered architecture to context-aware systems with the intention of hiding low-level sensing details. This technique eases extensibility since the client code has not to be modified anymore and it simplifies the reusability of hardware dependent sensing code due to the strict encapsulation. The last approach is to use a context server. This approach is used for permitting multiple client access to remote data sources. It extends the middleware based architecture by introducing an access managing remote component. Gathering sensor data is moved to context server to facilitate concurrent multiple access.

A number of research groups have developed context-aware systems. Xerox's Palo Alto Research Center has been working on pervasive computing applications since the 1980s. Carnegie Mellon University's Human Computer Interaction Institute (HCII) worked Aura project whose goal is to provide each user with an invisible halo of computing and information services that persists regardless of location [4]. HP's Cooltown project has developed technology future where people, place, and things are first class citizens of the connected world, a place where e-services meet the physical world. In this place, humans are mobile, devices and services are federated and context-aware, and everything has a web presence. The Massachusetts Institute of Technology (MIT) has a project called Oxygen [8]. The Oxygen system aims to bring an abundance of computation and communication to users through natural spoken and visual interfaces, making it easy for them to collaborate, access knowledge, and automate repetitive tasks.

However, these researches remained an experimental level on developing the prototype, because various context information and its achievement technologies were considered only for specific applications at specific platforms. Recently, against this problem, lively researches have been in progress for developing flexible middleware or server which can supply context-aware service infrastructure.

The architecture of the *Context Managing Framework* [9] presented by Korpipaa et al. use four main functional entities which comprise the context framework: the *context manager*, the *resource servers*, the *context recognition services* and the *application*. Whereas the resource servers and the context recognition services are distributed components, the context manager represents a centralized server managing a blackboard. It stores context data and provides this information to the client applications.

Another framework based on a layered architecture is built in the *Hydrogen* project [10]. Its context acquisition approach is specialized for mobile devices. While the availability of a centralized component is essential in the majority of existent distributed content-aware systems, the *Hydrogen* system tries to avoid this dependency. It distinguishes between a remote and a local context. The remote context is information another device knows about, the local context is knowledge our own device is aware of. When the devices are in physical proximity they are able to exchange these contexts in a peer-to-peer manner via WLAN, Bluetooth etc. This exchange of context information among client devices is called context sharing.

Context Toolkit [11] was developed as an intermediate dealing with context between a sensor and an application service. It collects context information from sensors, analyzes it, and delivers it to application services. The toolkits object-oriented API provides a super-class called *BaseObject* which offers generic communication abilities to ease the creation of own components.

Another extensible centralized middleware approach designed for context-aware mobile applications is a project called *CASS* (Context-awareness sub-structure) [12] presented by Fahy and Clarke. The middleware contains an *Interpreter*, a *ContextRetriever*, a *Rule Engine* and a *SensorListener*. The *SensorListener* listens for updates from sensors which are located on distributed sensor nodes. Then the gathered information is stored in the database by the *Interpreter*. The *ContextRetriever* is responsible for retrieving stored context data. Both of these classes may use the services of an interpreter.

RSCM (Reconfigurable Context-Sensitive Middleware for Pervasive Computing) [13] separates sensors and application services, and delivers necessary context information to applications through ad-hoc network. *RSCM* also collects context information, analyzes, and interpret it. When this context information satisfies conditions of registered application services, *RSCM* delivers it to the application services. It is the client-side to make a decision whether its application services should be executed in it or not. *RSCM* is based on *CORBA*.

The *SOCAM* (Service-oriented Context-Aware Middleware) project [14, 15] introduced by Gu et al. is another architecture for the building and the rapid prototyping of context-aware mobile services. It uses a central server called *context interpreter*, which gains context data through distributed context providers and offers it in mostly processed form to the clients. The context-aware mobile services are located on top of the architecture, thus, they make use of the different levels of context and adapt their behavior according to the current context. Also, it proposed a formal context model based on ontology using *OWL* to address issues about semantic representation, context reasoning, context classification and dependency.

GAIA [16] project also developed the middleware which can obtain and infer different kinds of context information from environments. It aims at supporting the development and execution of portable applications for active spaces. In *GAIA*, context is represented by 4-ary predicates written in DAML+OIL. The middleware provides the functions to infer high level context information from low level sensor information, and expect the context in advance by using the historical context information.

CoBrA (Context Broker Architecture) [7, 17] is an agent based architecture for supporting context-aware computing in intelligent spaces. Intelligent spaces are physical spaces that are populated with intelligent systems that provide pervasive computing services to users. Central to *CoBrA* is the presence of an intelligent *context broker* that maintains and manages a shared contextual model on the behalf of a community of agents. These agents can be applications hosted by mobile devices that a user carries or wears (e.g., cell phones, PDAs and headphones), services that are provided by devices in a room (e.g., projector service, light controller and room temperature controller) and web services that provide a web presence for people, places and things in the physical world. The *context broker* consists of four functional main components: the *Context Knowledge Base*, the *Context Inference Engine*, the *Context Acquisition Module* and the *Privacy Management Module*.

However, previous context-aware systems can not provide a complete solution for all the essential requirements in context-aware computing. In this paper, we propose a layered conceptual framework for developing a context-aware system based on previous related works and our further researches on context-awareness. This layered conceptual architecture separates context acquisition process from context manipulation process. Moreover, we describe how the context-aware system called CAMUS was developed based on this architecture. This paper also discusses several technical issues about development of context-aware applications for a network-based robotic system in which robots get various kinds of context information by communicating with a server and a server supports the management and processing of the context information for the robot services.

3 Layered Conceptual Structure for Context-Aware Systems

3.1 Context and Context Space

In order to develop computing systems that are context-aware, it is important to understand what is context and which constitutes context from an engineering perspective. Thus, a more precise definition of context must be offered for building context-aware systems.

Many researchers have attempted to define context by enumerating examples of contexts. Schilit et al. characterize context as a collection of information that describe the users in a context-aware system [4]. In their definition, contexts are the location of the user in a mobile distributed computing system. The identity of people and physical objects that are nearby the user and the states of devices that user interact with constitute contexts. While this definition characterizes the types of contexts, but it does not cover any other types of contexts that are useful for building context-aware systems such as the intentions of the users and the activities that the users are participating in.

Dey argues that a definition of context should not just be a list of information that describes users or the system because context is all about the whole situation that is relevant to an application and its set of users [11]. In his work, context is any information that can be used to characterize the situation of an entity. An entity is a person, or object that is considered relevant to the interaction between a user and an application, including the user and application themselves. For example, environmental attributes such as noise level, light intensity, temperature, and motion, system and device capabilities, available services, and user's intention could be context.

Context entities can be structured into three fundamental elements as many previous researchers have categorized [4, 18]. These fundamental elements are *user*, *computing entity*, and *environment*. This is widely adopted and used as the standard definition of context for context-aware computing.

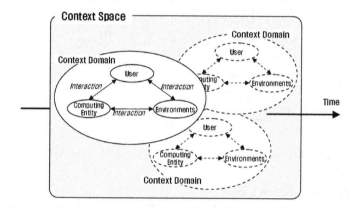

Fig. 2. The context space

In this paper, we propose the *context space* as a conceptual space that we want to represent for intelligent services in the context-aware computing environment as shown in Fig. 1. Three fundamental elements of context and their interactions constitute the *context space*. They can be grouped into one or more *context domains*. The *context domain* is a conceptual boundary of context information that a system or application developer is interested in and represent with a specific purpose. Context information in the *context space* is used for intelligent services by context-aware applications which is developed and executed based on context-aware system. In the *context space*, *time* is also an important and natural context for many applications. When the *computing entity*, *user* and *environment context* are changed as time goes on, we propose to add it as additional axis with the *context space*.

3.2 Context-Aware System

A system is context-aware if it uses context to provide relevant information and/or services to the user, where relevancy depends on the user's task [11]. Contexts which are acquired from various context sources are organized, represented and utilized in the context-aware system. Thus, a context-aware system should resolve issues about context organization, representation and context utilization. In Fig. 3, we describe the layered conceptual structure for context-aware systems and its main operations.

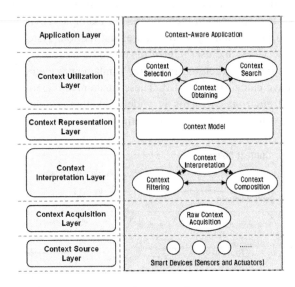

Fig. 3. The layered conceptual structure for context-aware systems

3.2.1 Context Acquisition Layer

Context acquisition layer concerns how to get the context information from various information sources. Raw context acquisition operation captures the raw data as the value of context features. Context sources include hardware sensors which provide explicit information given by the user or environment, and software services which can be used to collect computing context from information database or web-services. Depending on the type of context, one or multiple sensors may be employed. Data gathered from sensors can be very different.

3.2.2 Context Interpretation Layer

In this layer, context filtering, context composition and context interpretation operations are included. Some context information should be discarded after context acquisition or interpretation. Context filtering operation removes unnecessary context information and prevents duplicates. Context composition operation combines different contexts into another kind of context. Methods for composition are dependent on context types and the context model. Context interpretation operation obtains the semantics behind context information. Interpretation operation can be considered as high-level abstraction that aims at understanding an activity or revealing the purpose.

3.2.3 Context Representation Layer

After context acquisition and interpretation layer, contexts are stored and managed as the context model for further utilization. Context should be well organized into various data structures. The context storage maintains contexts and provides applications with shared context information under requests.

3.2.4 Context Utilization Layer

Context utilization concerns how to get relevant context information to context-aware applications. Different applications may expect to utilize different relevant contexts for their purpose.

Context obtaining operations obtains original context from the context model. There are mainly three methods to obtain context information: explicit query, polling, and event-driven. Context search operation search related context information that cannot be obtained directly by context obtaining operation. Context selection operation chooses only the relevant context information for application usage. Generally the selection could be made by setting a filter with matching conditions, and then detect incoming context accordingly.

4 The Context-Aware System: CAMUS

4.1 System Overview

Recently, Korea government has strategically promoted the development of a new concept of the network-based intelligent robot, which is called by URC. The main approach of URC is to distribute functional components through the network and to fully utilize external sensors and processing servers. The URC system requires not only the hardware infrastructure such as ubiquitous sensor networks and high-performance computing servers, but also the software infrastructure for context-aware applications development and execution. We developed CAMUS as the context-aware server framework for a network-based robotic system [5]. The functional architecture of CAMUS is shown in Fig. 4.

In the *service agent manager*, the sensing data from various sources such as physical sensors, applications and user commands are acquired and interpreted. After interpretation, sensing events are delivered to the upper layers in order to indicate the changes of contexts. The *service agent manager* supports the prerequisite for the proactive context-aware services. Also, it searches, invokes, and executes necessary services for each of *CAMUS* tasks.

The *event system* generates and manages events from physically distributed environments and is responsible for exchanging messages among *CAMUS* components. Above all, it delivers the events to context manager and task manager so that *CAMUS* tasks should be able to recognize the changes of situation and further update the existing context model.

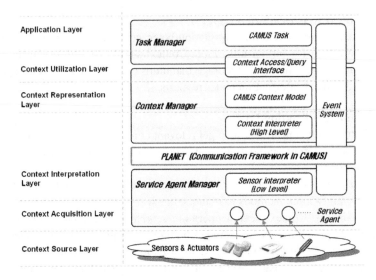

Fig. 4. The functional architecture of the CAMUS

The *context manager layer* infers implicit knowledge based on the context data which are delivered through events, and it also manages this inferred knowledge. When CAMUS tasks are executed, they refer to the context knowledge managed by this layer. Therefore, the context manager should be able to provide several main

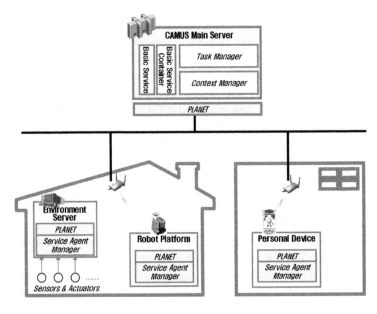

Fig. 5. System organization in the physical environment

functions for utilizing the context knowledge; (1) the context modeling and the knowledge representation, (2) the management of the context data, (3) the query for the context knowledge, and (4) the inference of the implicit knowledge, and so on.

The *task manager layer* executes each of tasks, manages and controls running tasks. Each task is executed based on the state transitions and the task ECA (Event-Condition-Action) rules. Events and Conditions are described for reflecting both the changes of environments and requirements of application and users. Action is services which are ready to act when these Event and Conditions are satisfied. Whenever this ECA task rule is fired, relevant context knowledge is referred through the context manager layer.

Under this functional architecture, *CAMUS* is mainly divided as a *CAMUS main server* and *service agent managers* based on the embedded types in the physical environment as shown in Fig. 5. The *CAMUS main server* communicates with *service agent managers* through the communication framework called *PLANET*.

4.2 Context Acquisition and Interpretation

The *service agent* is a software module performed as a proxy to connect various external sensors and devices to CAMUS. That is, it delivers information of the sensors in environment to the CAMUS main server. Also, it receives control commands from CAMUS main server, controls devices in the environment, and conducts applications.

The service agent is executed in the *service agent container* which is provided by the service agent manager. The service agent manger makes accesses to adequate service agents out of the currently running service agents using a searching function. Fig. 6 shows the architecture of the *service agent manager*.

The *service agent loader* manages the lifecycle of service agents by downloading, installing, starting, stopping and uninstalling them. These loaded service agents communicate with the CAMUS main server.

Information obtained by service agents is filtered or aggregated throughout the *sensor interpreter* as depicted in Fig. 7. This interpreted information will be delivered as a type of event to the CAMUS main server through the *event publisher*.

Short-term history queue (STHQ) in Fig. 7 is the temporal storage for generating new event through the context interpretation process or manipulating event through the context filtering and composition processes. STHQ is shared between all service agents that is managed by specific service agent manager. The entry which is written in STHQ are as follows:

- the CamusEvent entity: generated event
- timestamp: time that the event is generated
- delivered: whether the event is delivered or not

Event interceptor can be included in the sensor framework for making context information that is needed in the context-aware application. Event interceptor chain connects event interceptor in the sensor framework. Through this chain, many event interceptors can be joined to generate new event. Event operations of event interceptor in SHTQ are as follows:

- Event Canceling
- Event Transformation
- Event Generation

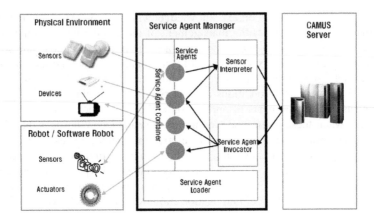

Fig. 6. Architecture of the service agent manager

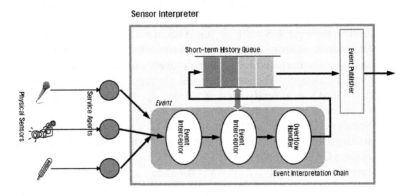

Fig. 7. The sensor interpreter

4.3   Representation of the Context Model

Context information in the CAMUS can be represented by using the Universal Data Model (UDM) and stored using RDB. As Fig. 8 shows, the UDM represents context information as nodes and associations between them. A node represents an entity such as person, place, task, service, etc. Every node has its unique ID and type. There are special nodes which have "valued" type (black circle in Fig. 8). These nodes can acquire values directly from class objects. Because the association starts and stops at a node, the association has a direction, defined by which node is start "From" (Fnode) and which node is designated "To" (Tnode). Fig. 8 shows an example of UDM representation.

The context model which is represented and managed in CAMUS includes the user context, environment context, and computing device context as shown in Fig. 2. User context includes user profile, user's task information, user preference, and so on. Environment context includes hierarchical location information, time, and so on. Computing device context includes information about available sensors and actuators.

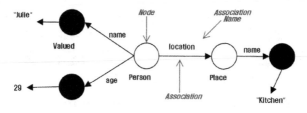

Fig. 8. An example of the UDM structure

4.4 Context Utilization in Task Manager

In the CAMUS, the context-aware application can be described by using the PLUE (Programming Language for Ubiquitous Environment) and is called task. The PLUE supports the state transition model and rule-based programming, which play a key role in presenting proactive, intelligent and invisible services to CAMUS users. That is, it enables application developers to easily write task states and the rules in each state. It is basically an extension of Java programming language, and in fact, its compiler is a pre-processor of the Java compiler.

The *Task* can be regarded as a set of work items which are required to be taken in a specific context or by the user's command in ubiquitous computing environment. Each work item, which is a unit of action, is described by the *Task Rules*. Task Rules is described by the convention of Event-Condition-Action (ECA). The action part arranges the operations of service agent to be executed when the incoming event meets conditions. The service agent is accessed by Task through *service agent manager*.

Task manager initiates individual tasks and manages on-going task processes. It also controls and coordinates tasks by managing the current state information of each task. It also supports context obtaining, search, and selection operation as described in Fig. 3.

4.5 Task Example – The Robot Greeting Task

To understand how to develop the task in CAMUS, we first illustrate the scenario happening using the intelligent service robot. Example task is the robot greeting task. The success scenario of this task is as follows:

- Step 1: A user arrives at home and goes to the living room.
- Step 2: RFID reader installed in the living room recognizes the user's entrance from outside of the home..
- Step 3: RFID reader deliveries entrance information of the user through the event channel in CAMUS.
- Step 4: CAMUS task orders robot to move to the user in the living room.
- Step 5: Robot deliveries event about navigation completed to the CAMUS task through the event channel.
- Step 6: Robot says the greeting message using TTS (Text to Speech) service.
- Step 7: Robot deliveries messages that are left from the user's family.
- Step 8: The CAMUS task orders the robot to move standby position and wait for user's command.

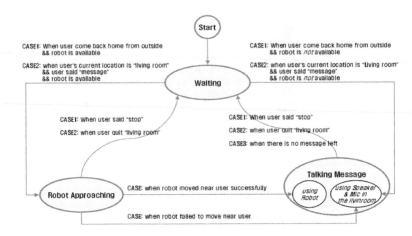

Fig. 9. State transition diagram of the robot greeting task

Fig. 9 shows the state transition diagram of the robot greeting task designed to implement the success scenario. The proposed task uses two events, *UserEntered* and *SpeechReceived*, in the CAMUS event channel. The *UserEntered* event generated by the RFID reader is related to the user's location, and the *SpeechReceived* event is related to the recognition of the user's voice commands. The example task consists of *Waiting*, *RobotApproaching*, and *TalkingMessage* states. As the user moves, task checks both current user's location and status of the robot, and then executes the related operation of services. For example, CASE 3 and 4 in the *waiting* state will be fired when the robot is not available in Fig. 9. At this time, the task will do message service using a speaker which is located in the living room, not in the robot. It shows very simple kind of service adaptation.

5   Field Test of CAMUS

The first step result of developing CAMUS was validated through URC Field Test during two months from Oct. 2005. to Dec. 2005. For URC Field Test, 64 households were selected from Seoul and its vicinity, where the broadband convergence network (BcN) was installed and 3 kinds of robots (Jupiter, Nettro and Roboid) were offered to them. Fig. 10 shows the URC field test structure for practical services of the CAMUS. Throughout the URC field test, we verified that the Hardware structure of a robot can be simplified by distributing robot's main internal functionalities (ex. speech recognition, and image recognition) to external servers through the network. And we also found that it is enabled to more effectively provide various kinds of services which are necessary for daily lives under the URC concept.

35 households out of 47 households' respondents showed a high degree of satisfaction, and they felt home monitoring, security, and autonomous cleaning services are most appropriate services for their daily lives. However, they felt a robot is worth only about 500 ~ 1,000 dollars with its performance, which implies that the robot market targeting households needs to make an effort to stabilize the robot price reasonably.

Fig. 11 shows the snapshot of Robot Greeting Task at the URC field test. The user's location is recognized by RFID sensor. When the user enters a living room from OUTWORLD, which means the outside of home, the robot moves to the user and provides the message service.

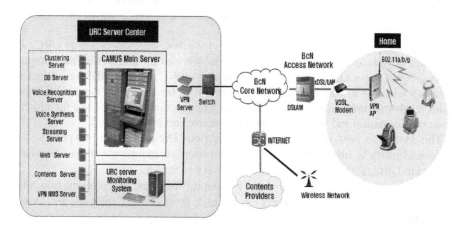

Fig. 10. System structure of the URC field test

Fig. 11. Snapshots of the robot greeting task at the URC field test

6 Conclusion

In the vision of ubiquitous computing environment, computer system understands their situational contexts and provides appropriate services to users. In this paper, we suggest a conceptual structure of context model for acquisition, representation, and utilization and describe how this conceptual structure is implemented in the service robot research area.

The new concept of the network based service robot, which is called as URC, was proposed to reduce robot costs and improve its services by distributing the computer processing of robots to a server over the network. We also describe introduced the CAMUS which is a context-aware server framework for the URC. The CAMUS provides a substructure which allows a robot to be connected with a server for utilizing context information and ultimately giving proactive services to users.

This system is now being applied to Ubiquitous Dream Exhibition sponsored by Korean Ministry of Information and Communication. As mentioned above, it was also applied to URC field test services, and it was very successful.

In the near future, we are planning to make up for some problems in rule inferences of the ontology based context model, high level context analysis, and system scalabilities.

References

1. Weiser, M.: The Computer of the 21st Century, Scientific American, Vol. 265. (1991) 66-75
2. Kindberg, T., Barton, J.: A Web-based nomadic computing system, Computer Networks, Vol. 35. No. 4. (2001)
3. Henricksen, K., Indulska, J., Rakotonirainy, A.: Modeling context information in pervasive computing systems, Proceedings of the First International Conference on Pervasive Computing, volume 2414 of Lecture Notes in Computer Science, Zurich, August (2002)
4. Schilit, W.N.: A System Architecture for Context-Aware Mobile Computing, Ph.D thesis, Columbia University (1995)
5. Kim, H., Cho., Y.J., Oh, S.R.: CAMUS - A middleware supporting context-aware services for network-based robots, IEEE Workshop on Advanced Robotics and Its Social Impacts, Nagoya, Japan (2005)
6. Baldauf, M., Dustdar, S., Rosenberg, F.: A Survey on Context-Aware Systems, Technical Report TUV-1841-2004-24, Technical University of Vienna (2004)
7. Chen, H.: An Intelligent Broker Architecture for Pervasive Context-Aware Systems. PhD thesis, University of Maryland, Baltimore County (2004)
8. Ranganathan, A., Campbell, R.H.: A Middleware for Context -Aware Agents in Ubiquitous Computing Environments, In ACM/IFIP/USENIX International Middleware Conference, Brazil, June (2003)
9. Korpipaa, P., Mantyjarvi, J., Kela, J., Keranen, H., Malm, E.-J. Managing context information in mobile devices. IEEE Pervasive Computing (2003)
10. Hofer, T., Schwinger, W., Pichler, M., Leonhartsberger, G., Altmann, J.: Context-awareness on mobile devices – the hydrogen approach. In Proceedings of the 36th Annual Hawaii International Conference on System Sciences (2002)

11. Dey, A.K., Abowd, G.D., Salber, D.: A Conceptual Framework and a Toolkit for Supporting the Rapid Prototyping of Context -Aware Applications, Anchor article of a special issue on Context -Aware Computing, Human-Computer Interaction (HCI) Journal, Vol. 16. (2001)
12. Fahy, P., Clarke, S.: CASS – a middleware for mobile context-aware applications. In Workshop on Context Awareness, MobiSys 2004 (2004)
13. Yau, S. S., Karim, F., Wang, Y.,Wang, B., Gupta, S.: Reconfigurable Context-Sensitive Middleware for Pervasive Computing, IEEE Pervasive Computing, joint special issue with IEEE Personal Communications, Vol. 1, No. 3 (2002)
14. Gu, T., Pung, H. K., Zhang, D. Q.: A middleware for building context-aware mobile services. In Proceedings of IEEE Vehicular Technology Conference (VTC), Milan, Italy (2004)
15. Wang, X.H., Zhang, D.Q., Gu, T., Pung, H.K.: Ontology based context modeling and reasoning using OWL, Proceedings of the Second IEEE Annual Conference on Pervasive Computing and Communications Workshops (2004)
16. Biegel, G., Cahill, V.: A Framework for Developing Mobile, Context-aware Applications, IEEE International Conference on Pervasive Computing and Communications (PerCom) (2004)
17. Chen, H., Finin, T., Joshi, A.: An ontology for context-aware pervasive computing environments, The Knowledge Engineering Review, Vol. 18. No.3 (2003)
18. Sun, J., Sauvola, J.: Towards a Conceptual Model for Context-Aware Adaptive Services, 4th International Conference on Parallel and Distributed Computing, Applications and Technologies, Chengdu, China (2003)

Music for My Mood: A Music Recommendation System Based on Context Reasoning

Jae Sik Lee [1] and Jin Chun Lee [2]

[1] Dept. of Management Information Systems
[2] Dept. of Business Administration,
Graduate School, Ajou University
San 5, Wonchun-Dong, Youngtong-Gu, Suwon 443-749, Korea
leejsk@ajou.ac.kr, giny777@empal.com

Abstract. The context-awareness has become one of the core technologies and the indispensable function for application services in ubiquitous computing environment. The task of using context data for inferring a user's situation is referred to as context reasoning. In this research, we incorporated the capability of context reasoning in a music recommendation system. Our proposed system contains such modules as Intention Module, Mood Module and Recommendation Module. The Intention Module performs context reasoning that infers whether a user wants to listen to music or not by using the environmental context data. The Mood Module determines the genre of the music suitable to the user's context. Finally, the Recommendation Module recommends the music to the user. Context reasoning is implemented using case-based reasoning.

Keywords: Context Reasoning, Context-Awareness, Music Recommendation System, Case-based Reasoning, Data Mining.

1 Introduction

Ubiquitous computing is a method of enhancing computer use by making many computers available throughout the physical environment, but making them effectively invisible to the user [28]. In ubiquitous computing environment, even though the user does not directly order the computer what to do, the computer should do what the user wants calmly of its own accord. To make it possible, the computer needs to aware the user's situation. We call this characteristic of ubiquitous computing environment 'context-awareness'.

In this research, we employ the concept of context-awareness in a recommendation system. Recommendation can be defined as "the process of utilizing the opinions of a community of customers to help individuals in that community more effectively identify content of interest from a potentially overwhelming set of choices" [19]. Generally, the products are recommended based on the demographic features of the customer, or based on an analysis of the past buying behavior of the customer. If the customer accesses an online bookstore, for example, through Internet, then we can recommend him/her appropriate books by analyzing the book data that he/she is

P. Havinga et al. (Eds.): EUROSSC 2006, LNCS 4272, pp. 190 – 203, 2006.

browsing. In ubiquitous computing environment, however, we have to capture the customer's intention before he/she actively reveals his/her interest, and recommend appropriate products. In this research, we propose a music recommendation system based on context reasoning. In other words, the recommendations are made inferring the user's intention and mood. No research on music recommendation of which we are aware has addressed this subject.

The rest of the paper is organized as follows: The next section provides a brief overview of the related work, i.e., context-awareness and reasoning, recommendation systems and case-based reasoning. Section 3 provides an overall structure of our proposed music recommendation system based on context reasoning. Section 4 describes the implementation process of our proposed system. Section 5 presents the performance evaluation of the proposed system. The final section provides concluding remarks and directions for further research.

2 Related Work

2.1 Context-Awareness and Reasoning

Context-awareness is to use information about the circumstances that the application is running in, to provide relevant information and/or services to the user [5]. The term 'context-aware' was introduced by Schilit and Theimer [21]. They defined 'context' through giving a number of examples of context-location, identities of nearby people and objects, and changes to those objects.

The term context, however, has been defined by many researchers in various ways. Schmidt et al. [22] defined it using three dimensions: physical environment, human factors and time. Benerecetti et al. [2] classified context into physical context and cultural context. Physical context is a set of features of the environment while cultural context includes user information, the social environment and beliefs. Dey and Abowd [7] defined context as "any information that can be used to characterize the situation of an entity, where an entity can be a person, place, physical or computational object that is considered relevant to the interaction between a user and an application, including the user and the application themselves". Dey [6] presented four types of context, i.e., location, identity, time and activity, and provided a framework for defining a context by giving values to these four types.

By sensing context information, context-aware applications can present context information to users, or modify their behavior according to changes in the environment [20]. Dey [6] defined context-aware application as "a system is context-aware if it uses context to provide relevant information and/or services to the user, where relevancy depends on the user's task". Three important context awareness behaviors are the presentation of information and services to a user, automatic execution of a service, and tagging of context to information for later retrieval [7].

Early investigations on context-aware application were carried out at the Olivetti Research Lab with development of the active badge system [24, 26], sensing locality of mobile users to adapt applications to people's whereabouts. Another research can be found at Xerox PARC with the ubiquitous computing experiment, from which a first general consideration of context-aware mobile computing emerged [22]. Dey et al. [8] developed a context toolkit to support rapid prototyping of certain types of context-aware applications. Many researchers have adopted this toolkit approach [11, 12], while others have been developing a middleware infrastructure [9, 15]. Some researchers adopted data mining techniques in context-aware applications. Kofod-Petersen and Aamodt [13] incorporated context information as cases in case-based reasoning (CBR) for user situation assessment in a context-aware mobile system. Kumar *et al.* [16] proposed a context enabled Multi-CBR approach. It consisted of two CBRs, i.e., user context CBR and product context CBR, for aiding the recommendation engine in retrieving appropriate information for e-commerce applications.

In order for future applications to become intelligent ubiquitous applications that are capable of context-awareness, one of the main challenges is the capability of recognizing the user's context. In other words, the applications should be capable of grasping the user's intention. For example, John is now in the department store. By a context-awareness system, such factors as person, place and time are collected and accordingly the products that are judged as appropriate for him, e.g., men's suits and neckties, will be recommended to him. However, these recommended products are useless for his situation because he came to the department store to buy a birthday present for his girl friend. This useless recommendation occurs because the context is judged using the superficial factors only. If he came here to buy a present for his girl friend, the products for women should be recommended. Therefore, by analyzing his past buying patterns and the kinds of shops he has been browsing today, we need to find out his intention. The research area related to this example is called 'context reasoning'. Context reasoning is defined as "deducing new and relevant information to the use of application(s) and user(s) from the various sources of context data" [18].

2.2 Recommendation Systems

The recommendation systems are to recommend items that users may be interested in based on their predefined preferences or access histories [4]. Many of the leading companies such as Amazon.com, Google, CDNOW, LA Times and eBay, are already using personalized recommendation systems to help their customers find products to purchase.

Research on recommendation systems is motivated by the need to cope with information overload and the lack of user knowledge in a specific domain. Researchers believe that user's needs and preferences can be converted into product selections using the appropriate algorithms and the knowledge embedded in the system. To date, a variety of recommendation systems have been developed for various items such as books, movies, news, web pages and music, using such techniques as collaborative filtering, content-based filtering, Bayesian network, association rules and case-based reasoning.

Ringo is a pioneer music recommendation system using collaborative filtering [23]. In Ringo, each user is requested to make ratings for music objects. These ratings constitute the personal profiles. For collaborative recommendation, only the ratings of the users whose profiles are similar to the target user are considered. Whether a music objects will be recommended is then based on the weighted average of the ratings considered.

Kuo and Shan [17] developed a content-based music recommendation system. In their system, the users' preferences are learned by mining the melody patterns, i.e., the pitch information, of the music they listened. Chen and Chen [4] proposed a music recommendation system that employed three recommendation mechanisms, i.e., content-based method, collaborative filtering and statistics-based method, for different users' needs. In their system, music objects are grouped according to the properties such as pitch, duration and loudness, and users are grouped according to their interests and behaviors derived from the access histories.

Celma *et al.* [3] proposed a music recommendation system called 'Foafing the Music'. The Friend of a Friend (FOAF) project is about creating a Web of machine-readable homepages describing people, the things they create and do, their interests and the links between them. The 'Foafing the Music' system uses this FOAF information and the Rich Site Summary that publishes new releases or artists' related news, for recommending music to a user, depending on his/her musical tastes.

2.3 Case-Based Reasoning

Case-based reasoning (CBR), a well-known artificial intelligence technique, has already proven its effectiveness in numerous domains. There are two fundamental concepts in CBR. The first one is that the same problems will often occur. The second one is that similar problems will have similar solutions. CBR is a method of solving a new problem by analyzing the solutions to previous, similar problems [14, 25, 27]. Since CBR can provide answers just using accumulated previous cases, i.e., case base, it can be applied to complicated and unstructured problems relatively easily. The most distinguished advantage of CBR is that it can learn continuously by just adding new cases to the case base.

As shown in Figure 1, CBR is typically described as a cyclical process comprising the four REs [1]: (1) REtrieve the most similar case or cases, (2) REuse the information and knowledge in that case to attempt to solve the problem, (3) REvise the proposed solution if necessary, and (4) REtain the parts of this experience likely to be useful for future problem solving.

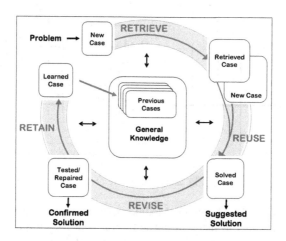

Fig. 1. Problem Solving Cycle of CBR

3 M³: A Music Recommendation System Based on Context Reasoning

In this section, we describe the structure and the components of the proposed system, we shall call M³ (Music for My Mood). Let us think about the following scenario:

> "Mira is a white-collar worker at her early 30s who enjoys listening to music. It is Friday and has been raining from morning, and it is chilly for a day in May. When she enters her room after returning home from work, her audio system is turned on automatically and plays Kenny G's saxophone number that she usually listened on a day like today."

In the above scenario, the music is selected by M³. The M³ consists of three layers, i.e., Interface Layer, Application Layer and Repository Layer as depicted in Figure 2. The role of Interface Layer is to collect environmental context data and to deliver the recommended music to the user. The Repository Layer is in charge of storing and managing relevant data. The Context DB stores the past environmental context data. The Listening History DB stores the list of music the user listened in the past. The Music DB contains the sound sources. The Application Layer infers the user's context and selects the music to recommend for the user. The Application Layer consists of three modules, i.e., Intention Module, Mood Module and Recommendation Module.

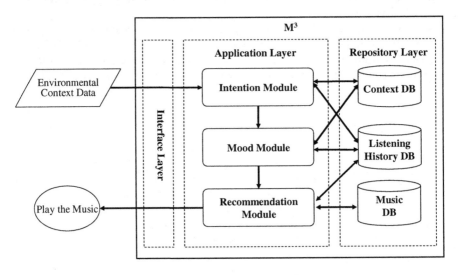

Fig. 2. The Structure of M³

1) Intention Module: The Intention Module performs context reasoning that infers whether a user wants to listen to music or not. It utilizes the user's past listening history and the corresponding environmental context data.

2) Mood Module: The Mood Module determines the style of music, i.e., slow or fast, suitable to the user's context. This module also utilizes the user's past listening history and the corresponding environmental context data.

3) Recommendation Module: The Recommendation Module recommends the music suitable to the user's mood to the user.

The M^3 recommends music to the user in the following steps:

Step 1: (Sensing the Context) Once the M^3 perceives the presence of the user, the Interface Layer collects the environmental context data and delivers it to the Intention Module. In our research, the environmental context data includes season, month, a day of the week, weather and temperature.

Step 2: (Inferring the Listening Intention) After receiving the environmental context data from the Interface Layer, the Intention Module infers whether the user wants to listen to music or not using the received context data and his/her listening history. If the Intention Module judges that the user wants to listen to music, then it delivers the context data to the Mood Module. Otherwise, the recommendation process stops. In this research, we employed case-based reasoning (CBR) as inference technique.

Step 3: (Inferring the User's Mood) The Mood Module infers the style of music suitable to the user using the environmental context data and his/her listening history. We employed CBR as inference technique. The inferred user's mood is delivered to the Recommendation Module.

Step 4: (Recommending the Music) The Recommendation Module generates the list of music suitable to the user's mood and delivers it to the audio system through the Interface Layer.

4   Implementation of M^3

4.1   Data Description

The data used in this research are Listening History data set and Weather data set. The Listening History data set was obtained from a streaming music service company in Korea, and it contains the list of songs and their genres listened by a certain customer, for example, 'Mira' for 6 months, or 171 days. The number of records in the Listening History data set is 14,373. The Weather data set was obtained from the Weather Bureau, and it contains the data such as season, month, a day of the week, atmospheric conditions and lowest, highest and average temperatures. These data sets are divided into three data sets as in Figure 3. Training data set is used to train the M^3 system, validation data set is used to optimize the system, and test data set is used to check the performance of the resulting system.

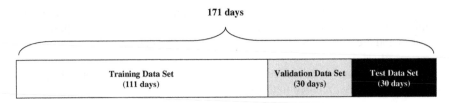

Fig. 3. Configuration of Model Data Sets

4.2 Implementation of Intention Module

In order to construct the Intention Module, we created the target feature 'WantMusic' and appended it to the Weather data set. If the user listened to music in a certain day, then the value of 'WantMusic' is 'Yes'. Otherwise, the value of 'WantMusic' is 'No'. Among the total of 171 days, Mira listened to music for 144 days, and did not listen to music for 27 days. Table 1 shows the proportion of the 'WantMusic' values.

Table 1. Proportion of the Feature 'WantMusic'

WantMusic	Number of Days	Proportion
Yes	144	84.2%
No	27	15.8%

For implementing the Intention Module, 7 features are used for input as shown in Table 2.

Table 2. Input Features for the Intention Module

Feature Name	Description	Type
Season	Spring, Summer, Fall, Winter	Categorical
Month	January, February, March, April, May, June, July, August, September, October, November, December	Categorical
Weekday	Monday, Tuesday, Wednesday, Thursday, Friday, Saturday, Sunday	Categorical
Weather	Sunny, Partly Cloudy, Mostly Cloudy, Cloudy, Rainy, Snow	Categorical
Avg_Temp.	Average temperature during a day (Unit: Celsius)	Numeric
High_Temp	Highest temperature during a day (Unit: Celsius)	Numeric
Low_Temp	Lowest temperature during a day (Unit: Celsius)	Numeric

In order to construct the Intention Module using CBR, we need to define the similarity function that will be used to find the old cases similar to the new case. The similarity between the cases N and C is calculated using the function presented in Equation (1).

$$Similarity\,(N,C) = \frac{\sum_{i=1}^{n} f(N_i, C_i) \times W_i}{\sum_{i=1}^{n} W_i} \tag{1}$$

N_i: the ith feature value of the new case.

C_i: the ith feature value of the old case.

n: the number of features.

$f(N_i, C_i)$: the distance function between N_i and C_i.

W_i: the weight of feature i.

The distance function in Equation (1) is to measure the similarity score of the features between two cases. For numeric features, Equation (2) is used. For categorical features, a partial matching scheme is devised using the domain knowledge as shown in Table 3.

$$f(N_i, C_i) = \begin{cases} 1-d & \text{if } 0 \le d \le 1 \\ 0 & \text{if } d > 1 \end{cases} \tag{2}$$

where

$$d = \frac{|N_i - C_i|}{Max - Min}$$

Max: maximum value among i^{th} feature values for all cases in case base.
Min: minimum value among i^{th} feature values for all cases in case base.

Table 3. Similarity Score Metrics for Categorical Features

Feature	Similarity Score Metrics				
Season					
	Old Case \ New Case	Spring	Summer	Fall	Winter
	Spring	1	0.2	0.5	0.2
	Summer	0.2	1	0.2	0
	Fall	0.5	0.2	1	0.2
	Winter	0.2	0	0.2	1
Month					
	Similarity Score	Condition			
	1	$N_i = C_i$			
	0.5	Distance Between N_i and C_i = 1 Month			
	0.2	Distance Between N_i and C_i = 2 Months			
	0	Otherwise			
...	...				
...	...				

CBR is basically based on the k-Nearest Neighbors algorithm. Therefore the performance of CBR model is affected by the value of k. Another factor that affects the performance of CBR is the weight vector of the input features. For implementing the best model, we first generated hundred weight vectors, and then experimented with the resulting CBR model by varying the value of k as 1, 3, 5 and so forth for each weight vector. As a result, the best CBR model for validation data set was obtained when $k = 3$ and the accuracy of the model was 93.3%.

4.3 Implementation of Mood Module

For implementing the Mood Module, we first classified the user's mood into three categories, i.e., 'Slow Music', 'Any Music' and 'Fast Music':

- Slow Music: The user likes to listen to the soft, sweet and calm music. The music genres that belong to this category are 'Ballad' and 'R&B'.

- Any Music: The user does not have any preferred genre.

- Fast Music: The user likes to listen to the cheerful, upbeat, and fast-paced music. The music genres that belong to this category are 'Rock/Metal' and 'Dance'.

The mood of the user in a certain day 'd' is determined by the Mood Score $M(d)$ calculated using Equation (3).

$$M(d) = \frac{\sum_i n_{id} \times G_i}{L_d} , \quad d = 1, \cdots, K \qquad (3)$$

where $i = \{Ballad, R\&B, Rock/Metal, Dance\}$

n_{id} : the number of listened music in the genre i on the day d.

G_i : the score of the genre i.

$$G_i = \begin{cases} 0.1 & if\ i = 'Ballad' \\ 0.3 & if\ i = 'R\&B' \\ 0.7 & if\ i = 'Rock/Metal' \\ 0.9 & if\ i = 'Dance' \end{cases}$$

L_d : the number of listened music on the day d.

K : the number of days.

Genre Score G_i is close to 0 as the music style is slow and 1 as the music style is fast. If, for example, the user listened to 10 songs in a certain day, say 2006/05/15, and 5 songs among them are 'Ballad', 3 songs are 'R&B', and 2 songs are 'Rock', then $M(2006/05/15)$ is calculated as 0.28 (=((5×0.1)+(3×0.3)+(2×0.7))/10) by Equation (3). Therefore, because $M(2006/05/15)$ is close to 0, we can say that in the environmental context of 2006/05/15, the user listened mostly to the soft music.

The user's mood is determined using the conditions defined in Table 4. $\overline{M}$ is the average value of all $M(d)$ s. If $M(2006/05/15)$ is 0.28 and $\overline{M}$ is 0.5, then the user's mood is determined as 'Slow Music'.

Table 4. Determination of User's Mood

User's Mood	Condition
Slow Music	$M(d) - \overline{M} <= -0.05$
Any Music	$-0.05 < M(d) - \overline{M} < 0.05$
Fast Music	$M(d) - \overline{M} >= 0.05$

Table 5 shows the proportion of Mira's mood in our data.

Table 5. Proportion of Mira's Mood

User's Mood	Number of Cases	Proportion
Slow Music	26	18%
Any Music	92	64%
Fast Music	26	18%

The Mood Module was also constructed using CBR. The best CBR model was found by the same manner as we did for the Intention Module. As a result, the best CBR model for validation data set was obtained when k = 5 and the accuracy of the model was 63.3%.

4.4 Implementation of Recommendation Module

The Recommendation Module recommends music to the user suitable to his/her mood. After the user's mood is determined, the top 15 songs (the number of songs normally contained in one CD) in the corresponding genres that the user listened most in the last one week are recommended. Table 6 shows the corresponding genres for each user's mood.

Table 6. Corresponding Genres for User's Mood

User's Mood	Music Genres
Slow Music	Ballad, R&B
Any Music	Ballad, R&B, Rock/Metal, Dance
Fast Music	Rock/Metal, Dance

For example, if Mira's mood is inferred as 'Slow Music', then the Recommendation Module recommends the top 15 songs in the 'Ballad' and 'R&B' genres that she listened most in the last one week.

5   Performance Evaluation of M^3

In order to evaluate the performance of M^3, we implemented two comparative systems. The first system is Con_CBR (meaning conventional CBR) that employs CBR technique and makes recommendations using the other users' listening histories stored in the case base. The second system is M^3-C (meaning M^3 minus context reasoning) that makes recommendations without making use of context data. All three systems were implemented using Microsoft Visual Basic 6.0 on PC. The comparison among Con_CBR, M^3-C and M^3 is presented in Figure 4.

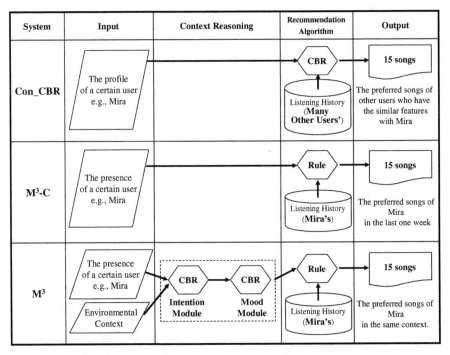

Fig. 4. Comparison among Con_CBR, M^3-C and M^3

For building Con_CBR, we obtained a separate data set consisting of 660 users. Among them, the data of 500 users is used as the training data set, i.e., the case base, and the data of remaining 160 users is used as the validation data set. The same test data set is used for evaluating the performances of Con_CBR, M^3-C and M^3. As input features for Con_CBR, we used such features as 'Age', 'Gender', 'Residence Area', 'Number of songs listened in the last one month' and so forth. The best Con_CBR was found by the same manner as we did in the previous section. The best Con_CBR was obtained when $k = 50$.

Now we describe the performance evaluation of M^3 compared to Con_CBR and M^3-C. Evaluation metrics are essential in order to judge the quality and performance of recommendation systems. There has been considerable research in the area of recommendation systems evaluation [10]. In this research, we use 'accuracy' as our choice of evaluation metric because it is the most commonly used and the easiest to interpret. The accuracy is calculated as follows:

$$Accuracy \;=\; \left(\sum_{i=1}^{N} \frac{|R_i \cap L_i|}{|R_i|} \times 100 \right) \bigg/ N \qquad (4)$$

where

N : the size of the test data set.

R_i : the set of recommended songs for the $i^{\,th}$ test case.

L_i : the set of actually listened songs in the $i^{\,th}$ test case.

Table 7 presents the accuracies of the three systems, i.e., Con-CBR, M^3-C and M^3 on the test data set.

Table 7. Accuracies of the Three Systems on Test Data

System	Accuracy
Con_CBR	71.0%
M^3-C	73.6%
M^3	89.8%

As shown in Table 7, the proposed system M^3 outperforms the other systems. The accuracy of M^3 is 89.8% and it is 18.8% point higher than Con_CBR and 16.2% point higher than M^3-C. The fact that M^3 and M^3-C are better than Con_CBR suggests that we can make a good recommendation system using the user's own data alone. The fact that M^3 is better than M^3-C definitely shows that the accuracy of recommendation can be increased by utilizing the environmental context data.

6 Conclusion

The contribution of this research is the development of a new recommendation system framework that infers the user's context by context reasoning and utilizes the user's own history data alone. In order to show the practicality of the framework, we implemented a music recommendation system, we shall call M^3, that can be used for a smart home application in ubiquitous computing environment. The M^3 has two distinctive features. First, the user's context is inferred using a data mining technique. Secondly, the music recommendation is made using the user's own listening history alone.

The M^3 contains such modules as Intention Module, Mood Module and Recommendation Module. The Intention Module performs context reasoning that infers whether a user wants to listen to music or not by using the environmental context data. The Mood Module determines the genre of the music suitable to the user's context. These two modules were developed using the case-based reasoning technique. Finally, the Recommendation Module recommends the music to the user by making use of the predefined rules.

The performance of M^3 was evaluated with the two comparative systems, i.e., the one that makes recommendation using multi-users' listening history data, and the other that makes recommendation without utilizing the environmental context data. The accuracy of M^3 outperformed the two comparative systems: 18.8% point higher than the former and 16.2% higher than the latter. Therefore, we improved the accuracy of music recommendation by making use of the user's own listening history alone and by utilizing the environmental context data.

For further research, we plan to continue our study on the following issues: First, we will develop M^3's for various individuals and check if the higher accuracy is maintained. Secondly, we will endeavor to make a more elaborate model for the Recommendation Module. Before starting our research, we tried to collect the feature that indicates the time of day when the user listened to certain music. This feature can be a good predictor for inferring the user's mood. However, we could not obtain the data of this feature. Therefore finally, we need to collect more various features that represent environmental context. Then, the input features for M^3 should be determined by feature selection process.

Acknowledgments. This research is supported by the Ubiquitous Autonomic Computing and Network Project, the Ministry of Information and Communication (MIC), 21st Century Frontier R&D Program in Korea.

References

1. Aamodt, A., Plaza, E.: Case-based Reasoning: Fundamental Issues, Methodological Variations, and System Approaches. Artificial Intelligence Communication, Vol. 7. (1994) 39-59
2. Benerecetti, M., Bouquet, P., Bonifacio, M.: Distributed Context-Aware System. Human-Computer Interaction, Vol. 16. (2000) 213-228
3. Celma, O., Ramírez, M., Herrera, P.: Foafing the Music: A Music Recommendation System based on RSS Feeds and User Preferences. In: Proceedings: 6th International Conference on Music Information Retrieval, London, UK (2005)
4. Chen, H.C., Chen, A.P.: A Music Recommendation System Based on Music and User Grouping. Journal of Intelligent Information Systems, Vol.24. (2005) 113-132
5. Cuddy, S., Katchabaw, M., Lutfiyya, H.: Context-Aware Service Selection based on Dynamic and Static Service Attributes. Wireless and Mobile Computing, Networking and Communications, IEEE International Conference (2005)
6. Dey, A.K.: Understanding and Using Context. Personal and Ubiquitous Computing, Vol. 5. (2001) 4-7
7. Dey, A.K., Abowd, G.D.: Towards a Better Understanding of Context and Context-Awareness. In: Proceedings of CHI 2000 Workshop on the What, Who, Where, When, Why, and How of Context-Awareness, The Hague, Netherlands (2000) 1-6
8. Dey, A.K., Salber, D., Abowd, G.D.: A Conceptual Framework and a Toolkit for Supporting the Rapid Prototyping of Context-Aware Applications. Human-Computer Interaction, Vol. 16. (2001) 97-166

9. Harry, C., Finin, T., Joshi, A.: An Intelligent Broker for Context-Aware Systems. Ubicomp 2003, Seattle, Washington (2003)
10. Herlocker, J., Konstan, J., Tervin, L.G., Riedl, J.: Evaluating Collaborative Filtering Recommender Systems. ACM Transactions on Information Systems, Vol. 22. (2004) 5-53
11. Hong, J.: The Context Fabric. Berkeley, USA, http://guir.berkeley.edu/projects/confab/ (2003)
12. Jang, S., Woo, W.: Ubi-UCAM: A Unified Context-Aware Application Model. In: Proceedings of Context 2003, Stanford, CA, USA (2003)
13. Kofod-Petersen, A., Aamodt, A.: Case-based Situation Assessment in a Mobile Context-Aware Systems. Workshop on Artificial Intelligence for Mobile Systems (AIMS2003), Seattle, October, (2003)
14. Kolodner, J.L.: Case-based Reasoning. San Mateo, CA: Morgan Kaufman (1993)
15. Kumar, M., Shirazi, B.A., Das, S.K., Sung, B.Y., Levine, D., Singhal, M.: PICO: a Middleware Framework for Pervasive Computing. IEEE Pervasive Computing, Vol. 2. (2003) 72-79
16. Kumar, P., Gopalan, S., Sridhar, V.: Context Enabled Multi-CBR based Recommendation Engine for E-Commerce. In: Proceedings of IEEE International Conference on e-Business Engineering (ICEBE'05) (2005) 237-244
17. Kuo, F.F., Shan, M.K.: A Personalized Music Filtering System Based on Melody Style Classification. In: Proceedings of IEEE international Conference on Data Mining (2002) 649-652
18. Nurmi, P., Floreen, P.: Reasoning in Context-Aware Systems. http://www.cs.helsinki.fi/u/ptnurmi/papers/contextreasoning.pdf, Helsinki Institute for Information Technology, Position Paper (2004)
19. Resnick, P., Varian, H.R.: Recommender Systems. Communications of the ACM, Vol. 40. (1997) 56-58
20. Salber, D., Dey, A.K., Orr, R.J., Abowd, G.D.: Designing For Ubiquitous Computing: A Case Study in Context Sensing. GVU Technical Report GIT-GVU (1999) 99-129
21. Schilit, B., Theimer, M.: Disseminating Active Map Information to Mobile Hosts. IEEE Network, Vol. 8. (1994) pp. 22-32
22. Schmidt, A., Beigl, M., Gellersen, H.W.: There is More to Context Than Location. Computers and Graphics, Vol. 23. (1999) 893-901
23. Sharadanand, U., Maes, P.: Social Information Filtering: Algorithms for Automating 'Word of Mouth'. In: Proceedings of CHI'95 Conference on Human Factors in Computing Systems (1995) 210-217
24. Want, R., Hopper, A., Falcao, V., Gibbons, J.: The Active Badge Location System. ACM Transactions on Information Systems, Vol. 10. (1992) 91-102
25. Wang, H.C., Wang, H.S.: A Hybrid Expert System for Equipment Failure Analysis. Expert Systems with Applications, Vol. 28. (2005) 615-622
26. Ward, A., Jones, A., Hopper, A.: A New Location Technique for the Active Office. IEEE Personal Communications, Vol. 4. (1997) 42-47
27. Watson, I.: Applying Case-based Reasoning: Techniques for Enterprise System. San Francisco, CA: Morgan Kaufmann (1997)
28. Weiser, M.: Hot Topics: Ubiquitous Computing. IEEE Computer (1993)

WLAN Location-Aware Application Based on Accumulated Orientation Strength Algorithm[*]

I-En Liao[1], Kuo-Fong Kao[1,2], and Ke-An Chen[1]

[1] Dept. of Computer Science, National Chung Hsing University
[2] Dept. of Information Networking Technology, Hsiuping Institute of Technolog
Taichung, Taiwan
ieliao@cs.nchu.edu.tw

Abstract. With the increasing demand for location-aware services, the location determination system has emerged as an essential component of e-society, and due to the wide-spread deployment of wireless local area networks (WLAN), the location determination system based on WLAN has become a hot research area. However, the prediction accuracy remains a primary issue for the practicality of WLAN-based systems. This study proposes an accumulated orientation strength algorithm that can smartly utilize uncertain estimated orientation information to improve prediction accuracy. A sample application for course information retrieval is built to demonstrate the usefulness of location-aware services. The experimental results indicate the effectiveness of the proposed approach.

Keywords: wireless LAN, location determination, location-aware, received signal strength, accumulated orientation strength.

1 Introduction

Applications of location-aware services are increasing in recent years due to the advances on location determination technologies. The best-known location determination system is the Global Positioning System (GPS)[1], which is based on signals from satellites. Although a mature system, its satellite signals are easily blocked by the shelter. Alternatively, the received signal strength (RSS) of wireless LAN (WLAN) from an access point (AP) can be used in indoor environments to locate mobile users. Therefore, the WLAN-based method is an attractive solution because of the popularity and low cost of WLAN systems.

WLAN location determination systems typically work in two phases, off-line training and on-line location determination. During the off-line phase, the system gathers RSS levels using a mobile device over a pre-defined area, and uses them to construct the radio-map. In the on-line phase, the real time signal received at the mobile device is compared with the radio-map, and a suitable match location is returned as the estimated location. The overall system performance

---

[*] This research was partially supported by the National Science Council, Taiwan, under contract no. NSC 95-2221-E-005-114.

P. Havinga et al. (Eds.): EUROSSC 2006, LNCS 4272, pp. 204–217, 2006.

depends mostly on the algorithm that determines the appropriate location in the second phase. Several published studies employ different algorithms for the second phase, including Bayesian[2,3,4,5,6], k-nearest neighbor[7,8,9,10] and neural network methods[11].

However, the WLAN channels are generally noisy, and the RSS is easily influenced by many complex factors, including movement by people around the device, neighboring Bluetooth devices, doors opening or closing, the device orientation and even the dust in the air. Good accuracy is not easy to obtain in systems based on RSS with such variant contributing factors. This study adopts the user's orientation to improve accuracy, because the mobile user orientation always causes significant RSS variance. Moreover, user's orientation can be detected by tracking the RSS variance without using additional hardware.

The effect of user orientation on RSS has been examined[8,12,13], but the feasible application using orientation information in location determination systems has not been addressed. Significantly the orientation information should be used with care, because the predicted orientation may be wrong, and wrong orientation information may produce highly inaccurate estimated location. This paper proposes a model that carefully uses the orientation information to increase the overall location prediction accuracy. A sample application for retrieving course information is also built to demonstrate the usefulness of location-aware services. This application can retrieve course web pages based on the current position, which is predicted by the proposed system, of the mobile user.

The rest of this study is organized as follows. Section 2 summarizes the location determination methods and discusses the impact of user orientation on RSS. Section 3 describes the experimental testbed. Section 4 presents algorithms for improving accuracy by using orientation information. Section 5 introduces the proposed location-aware application. Section 6 describes the simulation setup and results. Conclusions are finally drawn in Section 7.

2   Related Work

Many studies have focused on positioning systems using radio frequency(RF) methods, which can be classified into two categories depending on whether they are assisted by dedicated hardware. The systems in the first category, such as the Active Bat[14,15], Cricket[16], PinPoint[17] and recent RFID systems[18,19], use specialized hardware. The systems in the second category, such as the RADAR system developed by Microsoft[7,8], Nibble from UCLA[3], the CMU serial algorithm from Carnegie Mellon University[10], Rover and Horus system from University of Maryland[2,6,20], Rice University's Robotics-based method[4,5] and the calibration-free technique from the University of Twente[21] use the underlying wireless network without extra hardware. Some studies[22,23,24] have surveyed the development of there positioning systems. However, those studies are not discussed herein due to the space limitation.

The impact of the user orientation on the received signal strength may vary due to the direction of a mobile device's antenna, or the component of mobile device becoming a reflector of wireless signal. Some studies[25,26,27] have referred to this phenomenon as radio irregularity. Additionally, the user's body absorbs the wireless signal because the human body consists of 70% water. The resonance frequency of water is 2.4G, which is the same as that of the wireless signal.

Howard et al.[12] performed some interesting experiments on the effects of orientation on RSS. They recorded the signal strength in 48 hours using one robot, with people moving in the corridors and offices, or opening and closing doors. In their experiment, the robot stayed put without movement, the variations of RSS were found to be confined to around 10dB. The signal strength of a robot with 360° orientation rotation was also recorded, indicating a variation range of about 20dB. The effect without user presence is called the impact of mobile device orientation.

Howard et al.'s study used a robot, thus eliminating the effect of the human body. Conversely, Kaemarungsi and Krishnamurthy[13] focused on the influence of the user's presence, and found that user's body significantly affects RSS distribution. The standard deviation increased from approximately 0.68dBm to 3.00dBm when the user was present. The authors also performed the measurements at four orientations (facing North, West, South, and East). The experimental results demonstrate that the user's orientation has a significant effect on RSS. In one of their experiments, the sample mean of the RSS was lowest at −59.05dBm when the user faced south, and highest at −49.73dBm when the user faced west. In another experiment, no signal was received at a particular location when the user faced west, and the RSS at the other orientation was normal. The effect of the user's presence is termed the impact of user's orientation.

The wireless device is always in front of the user's eyes in real situations. Therefore the device changes its orientation to follow any change in orientation by the user. Hence, the RSS variation measurement of user's orientation always implies the impact of the orientation of mobile device.

3   Experimental Testbed

The experiment was performed on the seventh floor of the Science Building at the National ChungHsing University in Taiwan. The experimental area had dimensions of $62.5m \times 15.5m$ and includes 8 classrooms and the main hallway. The 239 measured points were spread on the floor about 5ft apart, except for some obstacles. Figure 1 displays the layout of the floor and the measured points with the upper right corner point numbered as $(1,1)$.

The experimental testbed had four APs, given by a_1, a_2, a_3, a_4, with two Buffalo WBR-G54 (11-32mW) APs and two DLink DI-724+ (20-30mW) APs. The original two APs, a_1 and a_2, were placed on the ceiling which was about three meters off the ground, and the extra two APs, a_3 and a_4, which were added for this research were attached to the air conditioner pipelines at about the same height as the previous APs. The four APs are marked with triangles in Fig. 1.

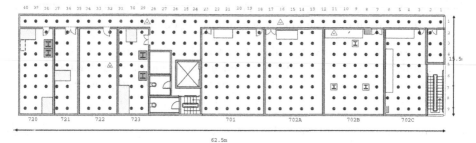

* : measured point   △ : access point

Fig. 1. Layout of experimental environment

A human operator collected data at each measured point using a IBM T30 notebook computer with a USB wireless Ethernet card. The notebook computer ran MS Windows 2000, and the wireless network driver was modified to record the RSS of the APs. At each measured point, the operator keyed in the coordinate and orientation, then started the data collection program. The operator remained about 1ft away from the notebook, and the data collecting program was set to detect the RSS automatically every one second. 300 records were collected at every orientation of all measured points, and all of the 286800 records were stored in a database table in an MS SQL Server 2000 database server. The RSS table schema comprised 7 attributes: RSS1, RSS2, RSS3, RSS4, Direction, X and Y.

The collected RSS data were distributed in the range between −20dBm and −88dBm. Any data point where no measurement was obtained was recorded as zero in the database. The whole data set was maintained in its original condition, and was not preprocessed. Preprocessing can distinguish between real and temporarily undetected measurements, thus improving the location estimation accuracy, but is not the focus of this study. To evaluate the algorithm performance, all recorded samples were partitioned into 70% training samples and 30% testing samples. These testing samples were fed to the basic Bayes classifiers and the proposed AOS algorithm, which is described in the next section.

4   Prediction Algorithm

4.1   Basic Bayesian Prediction Model

The Bayesian prediction method works in two phases, off-line training and on-line location determination. The system collects RSS data over a pre-defined area in the off-line phase. The RSS data are regarded as the observed data, and the positions are considered as the class labels. A large quantity of off-line phase data are applied to generate the RSS distribution over each position. In the on-line phase, the real time RSS detected on the mobile user is recorded and the position is unknown, meaning that the class label is unknown. By the Bayesian classifier, the posterior probability of a position can be calculated by giving the recorded signal in the on-line phase, and the position with the maximum probability is the predicted position. The following symbols are defined:

- $L = \{l_1, l_2, ..., l_n\}$ denotes the set of locations. Each location l_i, $1 \leq i \leq n$, is defined as (x_i, y_i), where x_i and y_i represent the coordinates of two-dimensional space. Let θ_i be the corresponding mobile user orientation at l_i and $\theta_i \subseteq \Theta = \{East, West, South, North\}$.
- $A = \{a_1, a_2, ..., a_p\}$ denotes the set of access points in the map, where a_k represents one access point, $1 \leq k \leq p$.
- SV denotes the set of signal vectors. For each vector $sv = (rss_{a_1}, rss_{a_2}, ..., rss_{a_p})$ in SV, rss_{a_k} represents the signal strength of the access point a_k, for $1 \leq k \leq p$.
- $Sample$ represents the collection of triplet data (sv, l, θ), where $sv \in SV$, $l \in L$, and $\theta \in \Theta$. The entire collection is divided into training and testing samples. The location component in the training sample is used to train the classifier, and that in the testing sample validates the estimated location.

The orientation component of a sample does not exist in systems which do not use the orientation. Therefore, Bayesian estimation model without user orientation information is first derived. The testing samples, given by $Sample_{ts}$, is then considered as the collection of data pair (sv_{ts}, l_{ts}), and the training samples, given by $Sample_{tr}$, is that of the data pair (sv_{tr}, l_{tr}). Although the orientation component is omitted, the training data may still be generated in various orientations.

In the determination phase, for each test data (sv_{ts}, l_{ts}) in $Sample_{ts}$, a location $\hat{l}$ is predicted for signal vector sv_{ts}. Define $f_B(sv_{ts})$ as the estimated location returned by the Bayesian location determination function $f_B(\cdot)$ upon receiving the signal vector sv_{ts}. This can be formulated as follows:

$$f_B(sv_{ts}) = \hat{l} \text{ such that } P(\hat{l}|sv_{ts}) > P(l_i|sv_{ts}), \forall i \ 1 \leq i \leq n \qquad (1)$$

Let $sv_{ts} = sv_{tr}$, then

$$P(l_i|sv_{ts}) = P(l_i|sv_{tr}) = \frac{P(sv_{tr}|l_i) \cdot P(l_i)}{P(sv_{tr})} \qquad (2)$$

The access points are always assumed to be independent, so that the problem of estimating the joint probability distribution becomes the problem of estimating the marginal probability distributions, given by:

$$P(sv_{tr}) = P((rss_{a_1}, rss_{a_2}, ..., rss_{a_p})) = \prod_{k=1}^{p} P(rss_{a_k})$$

$$P(sv_{tr}|l_i) = P((rss_{a_1}, rss_{a_2}, ..., rss_{a_p})|l_i) = \prod_{k=1}^{p} P(rss_{a_k}|l_i)$$

The above three formulas generate

$$P(l_i|sv_{tr}) = \frac{\prod_{k=1}^{p} P(rss_{a_k}|l_i) \cdot P(l_i)}{\prod_{k=1}^{p} P(rss_{a_k})} = (\prod_{k=1}^{p} \frac{P(rss_{a_k}|l_i)}{P(rss_{a_k})}) \cdot P(l_i) \qquad (3)$$

If $\hat{l}$ is equal to l_{ts}, then this procedure makes the exactly correct estimation. To discuss in further detail, the function $Correct(l_1, l_2, d)$ is defined as the procedure to examine whether the distance between l_1 and l_2 is smaller than or equal to d. Then $AC(f_B(\cdot), Sample_{ts}, d)$ is defined as the accuracy of location determination algorithm $f_B(\cdot)$ tested by $Sample_{ts}$ in error distance d. This function can be formulated as follows, and is later used to compute the accuracy.

$$AC(f_B(\cdot), Sample_{ts}, d) =$$

$$\frac{|\{t \,|\, Correct(f_B(sv_{ts}), l_{ts}, d) = true, \forall\, t = (sv_{ts}, l_{ts}) \in Sample_{ts}\}|}{|Sample_{ts}|} \qquad (4)$$

The above formulation is appropriate with slight modification in the condition involving orientation. In the training phase, the training sample adopts the triplet data format $(sv_{tr}, l_{tr}, \theta_{tr})$. In the testing phase, if the orientation information θ is available, then the Bayesian function at Eq. (1) is modified as $f_B(sv_{ts}, \theta)$, and Eqs. (2) and (3) are also modified accordingly.

The location is estimated using only the training samples with orientation θ when the orientation information θ is adopted. Such a classifier trained by special orientation samples is called an oriented classifier, and is denoted by $f_B^{\theta}(\cdot)$. For example, the classifier trained by the east orientation samples is called the east orientation classifier, and the estimated result is denoted as $f_B^{east}(sv_{ts})$. The result of $f_B(sv_{ts}, \theta)$ is equal to that of the orientation classifier $f_B^{\theta}(sv_{ts})$.

4.2   Using Orientation Information to Improve Accuracy

This section introduces the method for detecting orientation by tracking. The Accumulated Orientation Strength (AOS) algorithm is applied to determine whether the orientation information can be used. This algorithm is based on the assumption that the distribution of incorrect orientation estimates is not biased toward a specific direction.

The heuristic implemented in the proposed system only uses the RSS data and is based on the tracking of user's movement. Once the mobile user detects the signal vector sv_1 and $f_B(sv_1) = \hat{l}'_1$ and at next moment detects sv_2 and $f_B(sv_2) = \hat{l}'_2$, then the direction from $\hat{l}'_1$ to $\hat{l}'_2$ is defined as the moving vector of the user. If $\hat{l}'_1$ is equal to $\hat{l}'_2$, then the system guesses that the user stays at the same location without movement, and the vector is set to zero. Because the proposed system employs only four directions, it uses a four-dimensional variable $\omega = (East, West, South, North)$, called the orientation strength variable, to record the orientation strength of the mobile user on each direction. The projections of the vector are normalized and assigned as the value on each dimension, which is between 0 and 1, while the sum of all dimensions equals 1. For example, if the moving vector is toward east, then the orientation strength variable is $(1,0,0,0)$. If the moving vector is toward the northeast, then the orientation strength variable is $(0.5,0,0,0.5)$. The zero vector of the orientation strength is defined as $(0.25, 0.25, 0.25, 0.25)$. Figure 2 gives the detailed algorithm for computing orientation strength variable.

Input: location $\hat{l}'_1$ and $\hat{l}'_2$
Output: a four dimensional orientation strength variable $w = (e, w, s, n)$
Algorithm:

  Let moving vector $mv = \hat{l}'_2 - \hat{l}'_1$
  Get the projection of mv on X axis as mv_x
  Get the projection of mv on Y axis as mv_y
  if $mv_x = 0$ and $mv_y = 0$ then $w = (0.25, 0.25, 0.25, 0.25)$
  else

   Let $h = \frac{|mv_x|}{|mv_x|+|mv_y|}$ and $v = \frac{|mv_y|}{|mv_x|+|mv_y|}$
   // set east and west
   if $h = 0$ then $e = 0, w = 0$
   else if $mv_x > 0$ then $e = h, w = 0$
   else $e = 0, w = h$
   end if
   // set south and north
   if $v = 0$ then $s = 0, n = 0$
   else if $mv_y > 0$ then $s = 0, n = v$
   else $s = v, n = 0$
   end if

  end if

Fig. 2. Algorithm for computing orientation strength variable

The orientation strength variable cannot be used directly in the program, because it is calculated from the estimated location and may be incorrect. A wrong location estimation produces a false moving vector. If this information is applied directly, then the loss cannot be avoided. To solve this problem, multiple moving vectors are adopted instead of a single moving vector. Additionally, incorrect estimations of locations did not move in a fixed direction. For example, the predicted location is at the east of the real position at one moment, and at the west of the real position at the next moment. The errors seldom occur in the same direction of the real position. Hence, the wrong orientation estimation was assumed not to keep stay in a fixed direction continuously. The predicted locations of all testing samples were recorded to provide evidence for the above assumption. In our experiment, each of the non-boundary measured point was treated as the origin of a two-dimensional coordinate plane. The coordinate system was divided into four quadrants: the upper right area as the first quadrant; the upper left area as the second quadrant; the lower left area as the third quadrant, and the lower right area as the fourth quadrant. The testing samples with the same origin as the real positions were gathered, and the quadrants at which their predicted locations locate were collected. The number of incorrect estimations at each of the four quadrants was computed as the percentage of all incorrect estimations at each measured point. Finally, the percentages at each quadrant for all non-boundary measured points were averaged, and the result is listed in Table 1.

Table 1. Distribution of incorrect estimated locations

Area	First quadrant	Second quadrant	Third quadrant	Fourth quadrant
Probability	23%	20%	28%	29%

The distribution of incorrect estimations is not localized at a specific quadrant, although it is not uniformly distributed in four quadrants. Hence, incorrect orientation strength variables based on the estimated location does not bind to a specific direction. The prediction errors, if they occur, spread over all directions, and thus distinguish between correct and incorrect estimation. If a mobile user makes continuous movements toward a specific direction, and the system estimates the user's locations correctly, then the proposed system generates a sequence of orientation strength variables that indicate the same direction. This special pattern is different from the spreading pattern resulting from incorrect predictions. This subsection presents an AOS algorithm that efficiently distinguishes between these two patterns.

Assume a sequence of signal vectors $sv_1, sv_2, ...,$ and sv_n detected at time $1, 2, ...,$ and n, respectively. ω_i is defined as the orientation strength variable at time i, for $i = 1, 2, ..., n$. For $i \geq 2$, ω_i is computed from the estimated location $f_B(sv_{i-1})$ and $f_B(sv_i)$ as described in Sect. 4.1. The orientation strength variable at time 1 is set to $\omega_1 = (0.25, 0.25, 0.25, 0.25)$. Define aos_i as the accumulated orientation strength variable at time i, and set aos_1 to ω_1. The accumulated orientation strength variable at time i, $i \geq 2$ is defined as

$$aos_i = aos_{i-1} * r + \omega_i * (1 - r), \quad where \ 0 < r < 1 \tag{5}$$

For example, the accumulated orientation strength at time 2 is calculated as $aos_2 = aos_1 * r + \omega_2 * (1 - r)$, and that of time 3 is given by $aos_3 = aos_2 * r + \omega_3 * (1 - r)$. The accumulated orientation strength is computed iteratively from time 2 to time n step by step.

The weight parameter r denotes the retaining rate of the previous accumulated orientation strength. The higher the value of r, the smaller the effect of the new RSS. If the value of r is small, then the new arrival RSS dominates the accumulated orientation strength, resulting in system instability. Conversely, if the value of r is large, then the system reacts slowly to the changes in direction. Therefore, the weight parameter r is set to 0.5 in the proposed system. The orientation with maximum strength value among four directions is chosen as our estimated orientation, and this maximum value is assigned as the confidence of this orientation. Figure 3 shows the flowchart for this process.

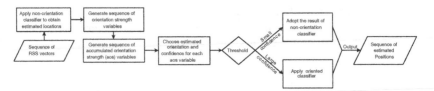

Fig. 3. Process of selecting oriented or non-orientation classifier

Input: a sequence of signal vectors $sv_1, sv_2, ..., sv_n$, the parameter r as the weight, and parameter h as the threshold value.
Output: a sequence of estimated locations $\hat{l}_1, \hat{l}_2, ..., \hat{l}_n$
Algorithm:
  declare variable l_p as the previous estimated location
  declare variable ω_p as the previous orientation strength variable
  declare variable aos as the accumulated orientation strength
  declare variable o as the estimated orientation
  declare variable c as the confidence of estimated orientation
  if receive the first signal vector then
      let $\hat{l}_1 = f_B(sv_1)$
      let $l_p = \hat{l}_1$ and $\omega_p = (0.25, 0.25, 0.25, 0.25)$
      let $aos = \omega_p$
  end if
  for each $sv_j, 2 \geq j \geq n$ do
      let $\hat{l'}_j = f_B(sv_j)$
      compute the orientation strength variable ω_j using l_p and $\hat{l'}_j$
      let $aos = aos * r + \omega_j * (1 - r)$
      choose the orientation with the maximum value in aos as o,
      and the value of the maximum orientation as confidence c
      if $c > h$ then let $\hat{l}_j = f_B^o(sv_j)$
      else let $\hat{l}_j = \hat{l'}_j$
      end if
      let $l_p = \hat{l}_j$
  end for

Fig. 4. Algorithm for computing accumulated orientation strength variables

If the orientation confidence is greater than the system threshold, then this orientation information is adopted. The system uses the specific orientation classifier to predict the location. Otherwise, the system still uses non-orientation classification results to avoid the risk of incorrect orientation estimation. Fig. 4 shows the accumulated orientation strength algorithm. The threshold value is the degree of tolerance to error estimation. A system with a high threshold value can tolerate a highly continuous error estimation. However, increasing the error tolerance reduces the probability of adopting the orientation information, and also reduces the value of benefit obtained from the orientation information.

This method can obtain efficiently the benefit from the true moving vector, and avoid the risk from the false moving vector, because the probability of error estimation at the same orientation is low. Even with many errors, those errors are not biased toward a specific orientation, making the orientation information indistinguishable. Hence the estimation errors rarely make the system choose the error orientation, thus avoiding the risk of using the wrong orientation information.

5 Location Aware Application

In our experiment, a course information retrieval system was created to demonstrate the usefulness of location-aware applications. The system uses a client-server architecture in which the server houses the location determination program and course information. The mobile device of a client can detect the signal strength of APs, and immediately sends the detected RSS to the server. Once the server receives the signal strength information, it executes the location prediction algorithm, and sends the predicted location and the course information for that location at that time to the client.

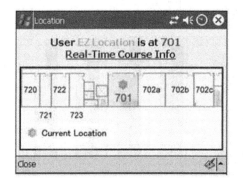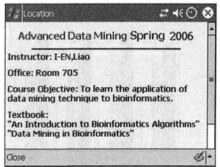

Fig. 5. Snapshots of location-aware course information application

Both PDA and notebook computer versions have been implemented for the client mobile devices. The predicted location of a client using a PDA is shown on the left-hand side of Fig. 5. Once the user clicked the "real time course info" link on the screen, the server retrieves the course information for the current location at the current time. A sample result is shown on the right-hand side of Fig. 5.

6 Performance Evaluation

The RSS data of four path types were collected to verify the performance of the proposed algorithm. These data are real RSS samples selected from the testing set according to the simulated paths. The first path was a straight walking path from (1,1) to (40,1). The second path started at (4,9), and ended at (19,7), with a change of direction at (3,9), (3,1), (23,1), (23,7). The third path simulated a user standing without movement. The fourth path simulated a user changing direction at each step. These four paths represent various real world situations. A successful system should be able to handle all these cases.

For each point in a simulated path, the simulator randomly chose an RSS sample from the testing set according to the coordinates and moving direction of that point. Each path was simulated 100 times. The resulting 100 simulated data sets were fed to classifiers, namely the non-orientation, match-direction, and AOS classifier. These classifiers were implemented using Weka library[28], which is an open source collection of machine learning algorithms for data mining tasks. The system performance can be evaluated from the average of prediction accuracy of 100 simulated data sets. In this simulation, the parameter weight r was set to 0.5, and the threshold parameter was set to 0.6. Figgure 6 shows the experimental results at error distances 1 and 2. The accuracy was computed using the Eq. 4 in Sect. 4.1.

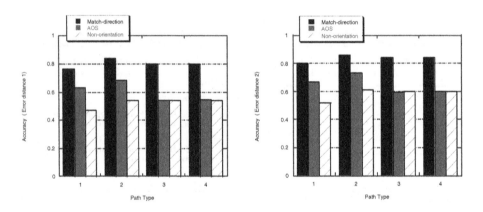

Fig. 6. Accuracies of different algorithms on path 1, 2, 3 and 4

The accuracy of the match-direction classifier can be viewed as the upper bound, and the accuracy of non-orientation classifier can be viewed as the baseline for comparison. The accuracies of AOS classifier on path 1 and 2 were larger than those of the non-orientation classifier. Conversely, the accuracies of AOS classifier on path 3 and 4 were almost the same as those of the baseline classifier. On path 3 and 4, the orientation estimation rarely remained in the same direction long enough regardless of the correctness of the orientation estimation. Hence, the proposed system abandons the opportunity to obtain benefit and acts like the non-orientation system. This behavior does not mean that the proposed algorithm fails. On the contrary, deliberately abandoning mined orientation information improves the stability of the system. The error distance factor does not significantly affect accuracy tendency. The above discussion is both applicable to the performance evaluations in error distances 1 and 2.

Another metric for measuring the merits of the proposed algorithm is the improvement ratio. The maximum improvement is defined as the accuracy difference between the non-orientation and match-direction classifiers. The aos-improvement is defined as the accuracy difference between the AOS and non-orientation classifiers. The improvement ratio of AOS classifier can then be defined as the ratio of the aos-improvement to the maximum improvement. Table 2 shows the improvement ratio of all tested paths in error distances 1 to 6.

Table 2. Improvement ratio for threshold value 0.6

Path/Distance	1	2	3	4	5	6
Path1	55.78%	54.73%	58.97%	55.59%	50.44%	47.91%
Path2	47.78%	50.00%	54.34%	52.04%	49.59%	48.48%
Path3	-0.01%	-0.02%	0.05%	0.08%	0.07%	0.07%
Path4	0.05%	-0.01%	0.01%	0.01%	0.05%	11.40%

The improvement ratios for paths 1 and 2 are all close to 50%. Conversely, the ratios are close to zero for paths 3 and 4. In summary, the proposed system can obtains benefit for straight walking paths, and avoids the loss for error estimation and other paths.

7   Conclusions and Future Work

Context-aware application is the critical component in modern e-society. This study demonstrates the location-aware course information retrieval system and proposes a novel scheme that utilizes the mobile user's orientation information to improve the prediction accuracy. The tracking-based method for detecting orientation information is implemented without additional hardware. In the proposed method, when the user is moving, the system can efficiently detect the user orientation from continuous estimated positions. Based on the assumption that the distribution of incorrect orientation estimations is not biased toward a specific direction, the system can stop applying orientation information when the estimated orientation is wrong. The system produces an accurate prediction position if it chooses the classifier according to the mined orientation information. To measure the performance, various paths were created, and the improvements of proposed systems were compared with the maximum improvement. The maximum improvement is defined as the difference in accuracy between the non-orientation and match-direction classifiers. When the user moves straight, the proposed system obtains the best performance and the improvement is about half of the maximum improvement. In the worst case, when the user changes direction at every step or does not move, the system maintains good performance by choosing the non-orientation classifier instead of oriented classifier. In conclusion, the system improves the overall accuracy by successfully avoiding the risk of using incorrect orientation information, and thus can gain the benefit of using correct orientation information.

The proposed technique can efficiently improve the accuracy of the wireless LAN location determination system. This approach is not limited to the Bayesian classifier, and can be applied in other classification algorithms. The authors are currently working to integrate other systems to improve the orientation information. The overall performance of the location determination system can be enhanced by the information from user profile or location profile. The placement of APs with maximum location prediction accuracy constraint is another potential future work. These further studies are expected to improve the accuracy. Wireless LAN location determination systems with increasing prediction capabilities will enable the development of context-aware applications in indoor environments, and improve the convenience of life in the future.

References

1. Logsdon, T.: Understanding the NAVSTAR: GPS, GIS, and IVHS (Second Edition). Kluwer Academic (1995)
2. Banerjee, S., Agarwal, S., Kamel, K., Kochut, A., Kommareddy, C., Nadeem, T., Thakkar, P., Trinh, B., Youssef, A., Youssef, M., Larsen, R.L., Shankar, A.U., Agrawala, A.: Rover: Scalable location-aware computing. Computer **35**(10) (2002) 46–53
3. Castro, P., Chiu, P., Kremenek, T., Muntz, R.R.: A probabilistic room location service for wireless networked environments. In: Proceedings of the 3rd international conference on Ubiquitous Computing, Springer-Verlag (2001) 18–34
4. Ladd, A.M., Bekris, K.E., Rudys, A., Kavraki, L.E., Wallach, D.S.: Robotics-based location sensing using wireless Ethernet. Wireless Networks **11**(1) (2005)
5. Ladd, A.M., Bekris, K.E., Rudys, A., Marceau, G., Kavraki, L.E., Wallach, D.S.: Robotics-based location sensing using wireless Ethernet. In: Proceedings of the Eighth ACM International Conference on Mobile Computing and Networking (MOBICOM), Atlanta, GA (2002)
6. Youssef, M.A., Agrawala, A., Shankar, A.U.: Wlan location determination via clustering and probability distributions. In: Proceedings of the First IEEE International Conference on Pervasive Computing and Communications, IEEE Computer Society (2003) 143
7. Bahl, P., Padmanabhan, V.N.: Enhancements to the RADAR user location and tracking system. Technical Report MSR-TR-2000-12, Microsoft Research (2000)
8. Bahl, P., Padmanabhan, V.N.: RADAR: An in-building RF-based user location and tracking system. In: INFOCOM (2). (2000) 775–784
9. Gwon, Y., Jain, R., Kawahara, T.: Robust indoor location estimation of stationary and mobile users. In: IEEE Infocom, Hong Kong. (2004)
10. Smailagic, A., Siewiorek, D.P., Anhalt, J., Kogan, D., Wang, Y.: Location sensing and privacy in a context aware computing environment. IEEE Wireless Communications **9**(5) (2002) 10–17
11. Battiti, R., Nhat, T.L., Villani, A.: Location-aware computing: a neural network model for determining location in wireless lans. Technical Report DIT-5, Informatica e Telecomunicazioni, University of Trento (2002)
12. Howard, A., Siddiqi, S., Sukhatme, G.S.: An experimental study of localization using wireless ethernet. In: Proceedings of the International Conference on Field and Service Robotics. (2003)

13. Kaemarungsi, K., Krishnamurthy, P.: Properties of indoor received signal strength for wlan location fingerprinting. In: Mobile and Ubiquitous Systems: Networking and Services, 2004. MOBIQUITOUS 2004. The First Annual International Conference. (2004)

14. Harter, A., Hopper, A., Steggles, P., Ward, A., Webster, P.: The anatomy of a context-aware application. In: Proceedings of the 5th annual ACM/IEEE international conference on Mobile computing and networking, ACM Press (1999) 59–68

15. Ward, A., Jones, A., Hopper, A.: A new location technique for the active office. IEEE Personal Communications 4(5) (1997) 42–47

16. Priyantha, N.B., Chakraborty, A., Balakrishnan, H.: The cricket location-support system. In: Proceedings of the 6th annual international conference on Mobile computing and networking, ACM Press (2000) 32–43

17. Werb, J., Lanzl, C.: Designing a positioning system for finding things and people indoors. IEEE Spectrum 35(9) (1998) 71–78

18. Ni, L.M., Liu, Y., Lau, Y.C., Patil, A.P.: Landmarc: indoor location sensing using active rfid. Wirel. Netw. 10(6) (2004) 701–710

19. Seshadri, V., Zaruba, G., Huber, M.: A bayesian sampling approach to in-door localization of wireless devices using received signal strength indication. In: PerCom 2005: Third IEEE International Conference on Pervasive Computing and Communications, 2005. (2005) 75– 84

20. Youssef, M., Agrawala, A.: The horus wlan location determination system. In: MobiSys '05: Proceedings of the 3rd international conference on Mobile systems, applications, and services, New York, NY, USA, ACM Press (2005) 205–218

21. Muthukrishnan, K., Meratnia, N., Koprinkov, G., Lijding, M.E.M., Havinga, P.J.M.: Svgopen conference guide: An overview. In: Proceedings of the 4th Annual Conference on Scalable Vector Graphics (SVG'05), Enschede, The Netherlands. (2005)

22. Chen, G., Kotz, D.: A survey of context-aware mobile computing research. Technical Report TR2000-381, Dept. of Computer Science, Dartmouth College (2000)

23. Hightower, J., Borriello, G.: A survey and taxonomy of location systems for ubiquitous computing. Technical Report UW-CSE 01-08-03, Dept. of Computer Science and Engineering, University of Washington (2001)

24. Muthukrishnan, K., Lijding, M.E.M., Havinga, P.J.M.: Towards smart surroundings: Enabling techniques and technologies for localization. In Strang, T., Linnhoff-Popien, C., eds.: 1st Int. Workshop on Location- and Context-Awareness (LoCA), Oberpfaffenhofen, Germany. Volume LNCS 3479., Berlin, Springer-Verlag (2005) 350–362

25. Woo, A., Tong, T., Culler, D.: Taming the underlying challenges of reliable multi-hop routing in sensor networks. In: SenSys '03: Proceedings of the 1st international conference on Embedded networked sensor systems, New York, NY, USA, ACM Press (2003) 14–27

26. Zhao, J., Govindan, R.: Understanding packet delivery performance in dense wireless sensor networks. In: SenSys '03: Proceedings of the 1st international conference on Embedded networked sensor systems, New York, NY, USA, ACM Press (2003) 1–13

27. Zhou, G., He, T., Krishnamurthy, S., Stankovic, J.A.: Impact of radio irregularity on wireless sensor networks. In: MobiSys '04: Proceedings of the 2nd international conference on Mobile systems, applications, and services, New York, NY, USA, ACM Press (2004) 125–138

28. Witten, I.H., Frank, E.: Data Mining: Practical machine learning tools and techniques, 2nd Edition. Morgan Kaufmann (2005)

Context Delivery in Ad Hoc Networks Using Enhanced Gossiping Algorithms*

Syarulnaziah Anawar, Lorcan Coyle, Simon Dobson, and Paddy Nixon

Systems Research Group
School of Computer Science & Informatics
University College Dublin, Belfield, Dublin 4, Ireland
Syarulnaziah.Anawar@ucd.ie

Abstract. The dissemination of context data across a pervasive environment has proven to be a difficult problem. Techniques using gossiping algorithms offer simplicity and flexibility but often result in poor performance with respect to timeliness of delivery and communication cost. In this ongoing-work, we present enhanced gossiping algorithms that aim to improve the efficiency of context data delivery in a decentralised manner using network and data-driven approaches.

1 Introduction

Context delivery is important to facilitate context-based interaction in pervasive computing environments. The dissemination of context information around *ad hoc* networks is particularly challenging due to scalability issues and to the dynamic characteristics of the networks. Different devices will likely have differing context information requirements to make appropriate adaptation to their behaviour. By distributed data throughout a network, each device has only a partial store of the global knowledge. Also, out-of-context information is often delivered to devices due to vague information requirement and representation. Improvements to existing distribution algorithms should attempt to minimise these problems.

Gossiping communication algorithms [1] can provide a robust and scalable solution to the problem of distributing context throughout an *ad hoc* network. Nodes (representing devices) maintain a table of other nodes known to them (their neighbours), and periodically select a node with which to exchange or gossip context data. Gossiping algorithms include a degree of unpredictability, but require far less in the way of guarantees about network structure, communications loss, and latency than other communications techniques. The high robustness offered by gossiping has utility for dealing with *ad hoc* arrangements of devices to be found in pervasive context-aware computing environments. However, heterogeneous node properties such as power, memory and connectivity make it unwise to blindly deliver massive volume of context data throughout

---

* This material is based on works supported by Science Foundation Ireland under Grant No. 04/RP1/I544.

P. Havinga et al. (Eds.): EUROSSC 2006, LNCS 4272, pp. 218–221, 2006.

a network of devices. The continuous range of probability and unpredictability, raises the need for more optimisation of each node's network utilisation and more subtle interpretation for context data. Therefore, it is important to locate interesting and meaningful relations of devices in order to increase efficiency of context data delivery using our improved gossiping implementation.

2   Improved Gossiping Techniques

The distinction between our approach and those described by Voulgaris et al [1] is the use of selective information dissemination and discovery techniques for determining the most relevant context data to gossip, and most relevant nodes to gossip to, in a purely decentralised manner. We propose algorithms that analyse characteristics of the network and build multiple logical networks (overlays) connecting nodes independently of the underlying physical network. The overlays construction are built using a combination of three approaches:

Network-Centric. Minimising the volume of superfluous context data that propagates across physical network boundaries by organising nodes into clusters.

Data-Centric. Semantically defining the context of interest for a node and transferring context data to those neighbouring nodes with mutual interest.

Query-Centric. Routing a node's request for a specific pieces of data to the most relevant neighbours.

The *network-centric* approach uses an analysis of the role of network structure and behaviour in controlling the amount of context information delivered across a wide area network. Since network structure directly affects gossiping statistical properties (i.e. number of propagated messages, number of hops), finding the best way to characterise and customise network topology is fundamental. To achieve this, we propose to integrate information about the physical organisation of nodes to create logical views of the network. This can be done by examining the network topology about a node to estimate the relative importance of its neighbouring nodes in communicating with the wider network. Nodes are organised into clusters so that they will communicate mainly with each other. These clusters can be interconnected in such a way that there is a reduced probability of sending context information across physical network boundaries, and minimising overall network communication overhead.

The *data-centric* approach takes into consideration each node's context data consumption requirements during gossiping. To determine a node's requirement, we propose to describe context data semantically [2]; providing context specification and precisely defining topics of interest for each node. By examining the topic *utility* (the frequency of topic occurrence) at its disposal, it is possible for a node to compile a list of topics that it is interested in based on degree of importance for a particular context. For example, considering a node that runs location-tracking systems, semantic representation involving *weather* context may be regarded as having low utility. Using the data-centric approach, nodes prefer to gossip with neighbouring nodes that have mutual interests.

The data-centric approach may have some disadvantages. The technique requires content examination to resolve the differences between two nodes and may have large traffic overhead. To overcome this, if a node only requires a single piece of data, the *query-centric* approach to gossiping makes it possible for a node to issue a query to be sent to a set of neighbouring nodes. The basic principle is that queries are forwarded to the neighbouring nodes with a high likelihood of returning an answer. To achieve the objective, neighbours will be prioritised based on degree of similarity between a source node and its neighbour by analysing the semantic relationship between nodes. Semantic link is computed based on meta-data that each return messages were tagged with using following metrics:

Knowledge value. Relationship between nodes may be one of the these types; full or partial relationship, super-type or sub-type relationship, or semantically unrelated. Each type will denotes a knowledge value indicating node's knowledge on specific topic.

Confidence value. Confidence value will denote node's satisfaction competency (i.e. number of returned answer, accuracy of returned answer) with respect to a local query history.

This meta-information is computed to determine the degree of similarity between nodes on a scale from 0 to 1. Nodes will be ranked based on this scale and in future gossiping interactions, queries will be only forwarded to top ranked node in the neighbour table. If a new prospective neighbour is found to have a higher rank than the least similar node among the existing neighbours, the neighbour table will be dynamically updated. If a piece of data is requested frequently , it will be favoured and the node will continuously update its content to ensure the freshness of the context data. This is an important feature of this approach in improving the timeliness of delivery of sought-after data.

The data-centric and query-centric approach may be broadened by gossiping queries for topics of interest to neighbouring nodes that have no interest in those topics themselves. As these queries find nodes that can satisfy them, paths through the network will emerge in a stigmergic manner [3]. The nodes connected by query paths will form semantic community that links nodes sharing similar interest. Moreover, these paths would be dynamic, fault-tolerant and de-centralised.

3   Conclusion

Gossiping provides a simple means of communicating context data across unstructured ad hoc networks. The need for smarter gossiping algorithms to overcome the limitations of current approaches is the key motivation of this work. This paper presents some novel enhancements to the existing gossiping algorithms that consider local network awareness about a node, and the individual data requirements of nodes. Our next step is to perform evaluations of our proposed algorithms using a dedicated gossiping evaluation framework.

References

1. Voulgaris, S., Jelasity, M., van Steen, M.: A robust and scalable peer-to-peer gossiping protocol. In: Proceedings of the 2nd International Workshop on Agents and Peer-to-Peer Computing (AP2PC03), Melbourne, Australia (2003)
2. Clear, A.K., Knox, S., Ye, J., Coyle, L., Dobson, S., Nixon, P.: Integrating multiple contexts and ontologies in a pervasive computing framework. In: Contexts and Ontologies: Theory, Practice and Applications, Riva Del Garda, Italy (2006) To appear.
3. Bonabeau, E., Dorigo, M., Theraulaz, G.: Swarm intelligence: from natural to artificial systems. Oxford University Press (1999)

An Attribute-Based Naming Architecture for Wireless Sensor Networks Using a Virtual Counterpart Overlay Network*

Eui-Hyun Jung[1], Yong-Pyo Kim[2], Yong-Jin Park[2], Seong-Yun Cho[1], and Su-Young Han[3]

[1] Dept. of Digital Media, Anyang University
708-113, Anyang 5-dong, Manan-Gu, Anyang City, Kyunggi-do, Korea
{jung, scho}@anyang.ac.kr
[2] Dept. of Electronic and Computer Engineering, Hanyang University
Haengdang-dong, Sungdong-Gu, Seoul, Korea
{ypkim, park}@hyuee.hanyang.ac.kr
[3] Dept. of Computer Science, Anyang University
Samseong-ri, Buleun-myeon, Ganghwa-gun, Incheon, Korea
syhan@anyang.ac.kr

Abstract. In Wireless Sensor Networks, the attribute-based naming using data-centric characteristics of sensor nodes has been attracted researchers' attention because it is impractical for sensor applications to communicate directly to an individual sensor node. Related researches proposed effective data abstraction scheme essential for a smart sensing, but they reveals several weaknesses of heavy code problem, function extensibility and interoperability with existing sensor network architectures. We propose an attribute-based naming architecture using an overlay network consisting of virtual counterparts that represent physical sensors. Proposed attribute-based naming architecture provides a higher data abstraction for sensing data with intuitive data manipulation operations which can be easily extended without any modification of physical sensor nodes.

Keywords: Wireless Sensor Networks, Overlay Network, Virtual Counterpart.

1 Introduction

Even though data gathering and reporting are performed by communication among sensors in Wireless Sensor Networks (WSNs), it is not desirable to communicate directly to an individual sensor because sensor applications may be more interested in specific sensing data such as temperature in the specific area of sensing fields, rather than individual sensor itself [1]. For this reason, researchers have been focusing on a new kind of communication, named data-centric communication [1][2]. Data-centric communication doesn't collect data directly from each node's raw data but allocate abstract sensing tasks to whole sensor network. This kind of communication approach enables sensor applications to consider the entire sensor network as an abstract data infrastructure which is essential for rapid application construction and smart sensing.

* Part of authors who works for this research are supported by BK21's research funding.

P. Havinga et al. (Eds.): EUROSSC 2006, LNCS 4272, pp. 222–225, 2006.

To achieve data-centric communication, most of all, attribute-based naming is required. Attribute-based naming is a structure where application can easily assign abstract tasks to sensor networks and collect answers without considering low level end-to-end data communication. There have been several researches on sensor middleware capable of processing attribute-based queries such as SINA [2], Cougar [3], and SensorWare [4]. Even though existing researches are valuable in that they showed importance of attribute-based naming and feasibility, these middleware-based approaches are too heavy for tiny sensor nodes and have a defect in query extensibility.

This paper proposes a new attribute-based naming architecture based on virtual counterpart overlay network that supports cooperation among sensors and data fusion by locating virtual counterparts of physical sensors in the sink node without any additional overhead on physical sensor network. Moreover, when a new kind of query is required, the proposed overlay network enables extensibility only by modifying the overlay network in the sink node.

2 Proposed Architecture

Concept of virtual counterpart is a structure in which physical objects in the real circumstance are mapped to virtual objects on the cyber space [5]. In this paper, we designed an overlay network that provides attribute-based naming by locating virtual counterparts corresponding to real sensors in the sink node (i.e. base station) as shown in Fig. 1. The plentiful resource of the sink node in WSNs makes this structure feasible.

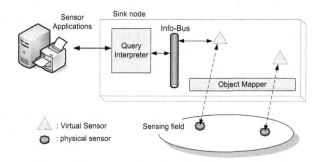

Fig. 1. The conceptual diagram describing the structure of the proposed architecture shows the relationship between a virtual sensor and a physical sensor

The architecture consists of Virtual Sensor, Object Mapper, Info-Bus, and Query Interpreter. A Virtual Sensor as a virtual counterpart for a physical sensor maintains collected data from a real sensor and responds to queries requested by sensor applications. Object Mapper has a role to deliver collected data from physical sensors to real sensors. To provide communication channel among Virtual Sensors, a Message Oriented Middleware (MOM), namely InfoBus, is designed. Query Interpreter performs a role to convert abstract queries requested by sensor applications into communication messages understandable by Virtual Sensors connected to InfoBus. It also gathers answers from Virtual Sensors and delivers fused answer to the sensor application sending query. Using this approach, sensor applications are able to get highly structured data without knowing about low level situation of physical sensor networks.

3 Operation of Attribute-Based Naming

The query syntax of proposed attribute-based naming architecture has a format as shown below.

Function [SensingAttr]* [Where Expr [LogicalOp Expr]*]

[] : optional, * : more than zero

Function in the first term indicates a function to manipulate data such as sum, average, or maximum value. Second term appoints target sensing data to obtain. Third term is used to specify conditional expression for filtering sensor nodes corresponding to a particular condition.

If we query the average temperature in the right bottom region of 100 m x 100 m sensing field, a sensor application on our architecture will send a query as follows.

AVERAGE Temperature Where Location OpIn Area (50,50,100,100)

For another sample, to get the maximum humidity from sensors with more than 10% of humidity in 100 m x 100 m sensing field, we will make a query as follows.

MAX Humidity Where Humidity OpGE 0.1

A query issued from sensor applications is delivered to Query Interpreter and it converts the query into a request message that can be transferred to Virtual Sensors through InfoBus. A query itself is converted to "ReqMessage" as a form of message packet and conditional terms in the query will be converted to the list of "QueryData". Conversion of the sample query related to maximum humidity is as shown in Fig. 2.

When this converted message is broadcasted to Virtual Sensors through Info-Bus, each Virtual Sensor compares it with its own Property and SensingData. Although all Virtual Sensors can listen to the "broadcast" message, only target Virtual Sensors will answer to the request message. For example, a thermal sensor located at the left top corner will drop the message if the request message targets humidity sensor or targets sensors located in right bottom region. To evaluate attribute-based naming query processing, twenty thermal sensors and ten humidity sensors are randomly located on 100 m x 100 m network topology with the J-Sim [6].

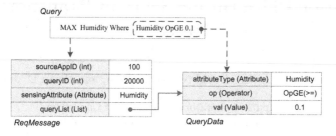

Fig. 2. Sample query is converted to a ReqMessage packet containing QueryData representing conditional terms

4   Conclusion

Data-centric communication scheme coincides very well with characteristics of sensor networks and attracts researchers for smart sensing. Most of current researches adopt a middleware approach to support data-centric communication and data abstraction of sensor networks. However, deploying a middleware on each tiny sensor node has some weaknesses especially about processing overhead and query extensibility. In this paper, an overlay network architecture consisting of virtual sensors is suggested to resolve these issues. A virtual sensor located in the sink node delegates a physical sensor and answers to abstract data queries from sensor applications. This structure can provide a highly customized query to sensor applications without any burden to the physical sensor network. If some applications want to add new query functions, the architecture can be easily extended by only modifying Query Interpreter. The proposed overlay network was simulated using the J-Sim simulator to verify its usefulness. The simulation result showed that proposed attribute-based naming performs requested functions properly. Currently, the proposed structure is being ported on the Tiny-OS.

References

1. Ian, F., and et al.: A Survey on Sensor Networks. IEEE Comm. Magazine, Vol. 40. Issue 8. (2002) 102–114
2. Chien-Chung, S. and et al.: Sensor Information Networking Architecture and Applications. IEEE Personal Communications. Vol. 8. Issue. 4. (2001) 52-59
3. Yong, Y. and et al.: The Cougar Approach to In-Network Query Processing in Sensor Networks. In SIGMOD. (2002).
4. Boulis, A. and et al.: Design and Implementation of a Framework for Efficient and Programmable Sensor Networks. Proc. of the First International Conference on Mobile Systems, Applications, and Services. (2003)
5. Kay, R. and et al.: Smart Identification Framework for Ubiquitous Computing Applications. Proc. of the PerCom'03 (2003).
6. J-Sim is a component-based, compositional simulation environment. AKA(Autonomous Component Architecture) : (http://www.J-Sim.org/).

A Sensor Platform for Sentient Transportation Research

Jonathan J. Davies, David N. Cottingham, and Brian D. Jones

Computer Laboratory, University of Cambridge,
15 JJ Thomson Avenue, Cambridge, CB3 0FD, UK
{jjd27, dnc25, bdj23}@cam.ac.uk

Abstract. This paper describes the creation of a vehicle-based sensor platform as part of research into sentient computing. We outline the challenges faced when building this platform and describe our techniques for overcoming them.

1 Introduction

The field of sentient computing is concerned with computers interpreting and reacting to sensor data in our every-day environments [1]. One such environment is that of transportation. We have prepared a vehicle-based, scalable sensor platform that is capable of sensing properties of the vehicle and its environment to provide context for intra- and extra-vehicular applications. Our goals for the platform included the ability for non-technical users to be able to use the vehicle without specialist knowledge, the collection of a large corpus of data and the creation of a general-purpose research platform.

2 Deployment

Cars are the most popular form of transportation on our roads and are thus an obvious subject for research into sentient transportation. However, building and operating an effective research platform requires significant in-vehicle space for its deployment, testing, and operation. Additionally, the vehicle should provide sufficiently unobtrusive trim in order to facilitate the installation of the equipment. We selected a Renault Kangoo van which meets these demands whilst maintaining a car-like exterior.

A vehicular environment is harsh, involving physical shocks, extremes of temperature and a lack of abundant power. This imposes limits on the computing equipment. We selected a Mini-ITX industrial-form-factor PC with a 1 GHz CPU consuming only 18 W, which performs comparably to an equivalent desktop PC (50 – 100 W), as a suitable choice for this environment. A 2.5" laptop hard drive was chosen to provide higher tolerance to vibration than conventional desktop drives.

P. Havinga et al. (Eds.): EUROSSC 2006, LNCS 4272, pp. 226–229, 2006.

2.1 Sensor Infrastructure

In order to facilitate research into context-aware applications, a platform is required which allows a wide range of sensors to be easily installed.

Sensors have a wide variety of power requirements and data rates. For example, a thermometer might emit a one-byte reading per second whereas the data rate from a digital video camera will be many orders of magnitude larger. Rather than attempting to provide a single data bus to meet these needs we elected to provide three separate communications buses. This permits the simple deployment of low data-rate sensors without precluding the use of sensors with more ambitious demands.

Low data-rate sensors are accommodated on a Controller Area Network (CAN) bus. Due to its rugged nature, similar technology is commonly used in vehicles for communication with built-in sensors. Interfacing a sensor with this bus does not require complex or high-speed hardware and is thus cheap.

Interaction with commodity devices such as GPS units and RFID readers is through a USB tree. This facility can also support ad-hoc attachment of personal devices. RS232-to-USB converters permit easy support for legacy serial devices. The length of the cabling required means that a USB 1.0 bus is more suitable than the higher-speed USB 2.0.

Higher data-rate devices are supported by a 100 Mbit Ethernet network. However, for such sensors, it is most preferable for them to perform as much processing as possible in hardware—such as MPEG encoding of video data—to relieve the on-board computer of this burden. The high speed nature of Ethernet means that interfacing is significantly more demanding than with the CAN bus.

At the time of writing, the platform integrates the following sensors:

- **CAN bus.** Thermometer, humidity sensor, barometer, tilt sensor, two-axis accelerometer, two-axis magnetometer;
- **USB.** GPS receiver, RFID reader, OBD-II interface;
- **Ethernet.** Digital video cameras.

The OBD-II interface is a standard supported by all modern vehicles for accessing data regarding the vehicle's operation. The specification of this standard supports a huge range of sensors but the subset which is available is vehicle-specific. In our vehicle, the available data include the engine speed (rpm), road speed, engine load (as a percentage of the peak available torque), air intake temperature, fuel rail pressure, intake manifold pressure and engine coolant temperature. Furthermore, the maximum rate at which data can be read from our OBD-II interface is limited to approximately 2 Hz. However, sensors such as the engine's velocity and load can vary substantially over the course of half a second. To minimise this problem as much as possible, we sample the sensors at adaptive rates sensitive to their operating characteristics.

2.2 External Communications

Data collected by our platform is not only useful for on-board context-aware applications, but also for those on other vehicles or fixed infrastructure. Hence, the vehicle needs to support communication with infrastructure providing ubiquitous coverage, and with more localised communication end-points. As a vehicle travels it will experience varying degrees of network connectivity; high-speed networks should be exploited when available.

To make use of WiFi hotspots, the vehicle has an IEEE 802.11b/g wireless LAN interface, which is also used as the primary means of communication when the vehicle is at base. In the future, inter-vehicular and vehicle-to-roadside communication will take place using the Dedicated Short Range Communications standard. To conduct testing of this technology in urban environments, we employ an IEEE 802.11a interface which operates in a similar manner. Meanwhile, a GPRS/UMTS interface provides a near-ubiquitous, low-bandwidth Internet connection.

2.3   Power

Vehicles cannot rely on any permanent connection to an external power source. This necessitates careful management of available power resources. In particular, it is important never to flatten the vehicle's main battery and to provide a stable power supply to the sensing platform. These goals were achieved through the use of an auxiliary battery, charged from the vehicle's alternator whilst the engine is running. A failsafe hardware power cut-off protects against draining the auxiliary battery in the event of the computer erroneously entering a stuck state.

For non-technical users to be able to use the vehicle, no specialist knowledge can be required to operate it. They cannot be expected to switch equipment on or off or monitor its state. We have designed the system to function in a fully-autonomous fashion, for all sensor data to be logged to permanent storage and uploaded automatically on return to base to enable post-processing.

3   Evaluation

Both technical and non-technical drivers have contributed to a corpus of sensor data for journeys in both urban and motorway environments. To date, we have collected over two and a half million data points, allowing a comprehensive picture to be built up of how the environment varies with location and time.

We have been able to easily install new sensors into our existing platform due to its scalability and modularity. In particular, we have found that utilising a dedicated sensor CAN bus has greatly simplified the communication interface between arbitrary sensors and the on-board computer.

Context-awareness plays an important role controlling the collection of data. One example is the use of camera data, where the frame rate is linked to the speed: when travelling at high speed on a motorway, there is little of interest, and hence the frame rate is decreased, whilst for low speeds (particularly in cities), a higher frame rate is desirable.

Our platform is being used for applications as diverse as driver expression inference [2], with a view to running real-time analysis on the driver's emotions;

automatically deriving digital road maps from GPS traces; and evaluating the performance of wireless technologies.

4   Related Work

An emerging example of utilising vehicles for mobile sensing is that of the collection of road traffic data [3] to permit authorities to better manage the transportation infrastructure and to permit navigation based on real-time congestion information.

Work has also been done in producing a platform for general collection of sensor data, such as the Instrumented Car [4] which is equipped with a variety of sensors obtaining data about the driver, the vehicle's emissions and its environment. Another related project is CarTel [5] which seeks to provide a mechanism for distributed sensor data to be queried and transferred. Our work builds upon and generalises these ideas into a multi-purpose vehicular platform for research into all aspects of sentient transportation, where the goal is accessibility, extensibility and scalability in the sensing platform.

5   Conclusion

This paper has described a vehicular sensing platform and has highlighted some of the challenges we experienced in its construction and their resolution. We hope that these experiences will help to inform the construction of similar vehicles in the future.

Acknowledgments

The authors would like to express their grateful thanks to Andy Hopper for his vision for, and support of, the project; and to Andrew Rice, Alastair Beresford and Robert Harle for their suggestions.

References

1. Hopper, A.: The Clifford Paterson lecture 1999: Sentient computing. Philosophical Transactions of the Royal Society **358**(1773) (2000) 2349–2358
2. Kaliouby, R.E., Robinson, P.: Mind reading machines: Automated inference of cognitive mental states from video. In: Proc. IEEE SMC. Volume 1. (2004) 682–688
3. Varshney, U.: Vehicular mobile commerce. IEEE Computer **37**(12) (2004) 116–118
4. Tate, J.E.: A novel research tool – presenting the highly instrumented car. Traffic Engineering and Control **46**(7) (2005) 262–265
5. Bychkovsky, V., Chen, K., Goraczko, M., Hu, H., Hull, B., Miu, A., Shih, E., Zhang, Y., Balakrishnan, H., Madden, S.: Data management in the CarTel mobile sensor computing system. In: Proc. ACM SIGMOD. (2006) 730–732

Attention-Based Information Composition for Multicontext-Aware Recommendation in Ubiquitous Computing

Sungrim Kim[1] and Joonhee Kwon[2]

[1] Department of Internet Information, Seoil College
49-3, Myonmok-dong, Jungrang-Ku, Seoul, Korea
srkim@seoil.ac.kr
[2] Department of Computer Science, Kyonggi University
San 94-6, Yiui-dong, Yeongtong-ku, Suwon-si, Kyonggi-do, Korea
kwonjh@kyonggi.ac.kr

Abstract. The recommender system that offers useful information to the users becomes more important in ubiquitous computing. Context-aware recommender systems are a core component in ubiquitous application. Because of the greater number of contexts, the total amount of information is larger in multicontext-aware environments. Multicontext therefore requires a more efficient recommendation method. We propose an attention-based information composition for multicontext-aware recommendation. We explain that the recommendation process would be comprised of three steps. Our proposed method is capable of recommending accurately and rapidly.

1 Introduction

We are facing a new generation of applications being characterized by the ubiquitous paradigm. In general, ubiquity offers new opportunities and challenges for applications in terms of time-, location-, device-aware and personalized services [3].

Context-aware recommender systems are a core component in ubiquitous computing. The goal of these systems is to proactively suggest adequate information the user may be interested in, taking into account the context at the moment of the interaction [1].

In multicontext-aware environments, the total amount of information can increase, due to the greater number of contexts. This requires accurate and rapid composition method of recommendations for multi-context. Previous approaches have typically supported only a single context. Although there have been some studies on multi-context, the composition of information over contexts is fixed at the design time and no attempt is made to dynamically compose via current contexts. However, the information a user can obtain from recommender systems will vary dynamically over contexts. In this paper, a new multicontext-aware recommendation method is proposed.

One of the main considerations of multicontext-aware recommendation is that information needs to be composed in accordance with the dynamic changes of each

P. Havinga et al. (Eds.): EUROSSC 2006, LNCS 4272, pp. 230–233, 2006.
© Springer-Verlag Berlin Heidelberg 2006

context. Because traditional approaches have taken into account only preferences, there are limitations to recommending while using dynamically changed attention. In this paper, not only user's preferences but also behavior as an attention factor is used in the dynamic composition process for a recommendation. Our proposed method recommends the more accurate information for the target context rapidly.

This paper is structured as follows: Section 2 will describe an overview of related work. Section 3 will discuss the proposed recommendation method. Finally, Section 4 will conclude the paper.

2 Related Work

It is important to incorporate the contextual information into the recommendation process. However, the traditional recommender systems have not taken into account the context when making recommendations. This seriously limits the relevance of the results.

Most previous context-aware recommendations are based on the match of a context to a user's preferences [3]. They work only with a specific single context. Although there are a few focusing on multi-context, the composition of information over contexts is fixed at the design time and no attempt is made to dynamically compose via current contexts. Moreover, they are not concerned about recommending information rapidly with the change of a user's contexts.

Other approaches to context-aware recommendations concentrate on recommending information rapidly in a ubiquitous computing environment. In [1], the context-aware cache tries to capture the information the user is most likely to need in future contexts. It makes for a more immediate retrieval and reduces the cost of retrieving information by doing most of the work automatically. In [4], the recommendation method locally stores the recommendation information that the user is likely to need in the near future based on user's context history in order to retrieve information rapidly. These approaches do not consider multi-context, however.

3 Attention-Based Information Composition for Multicontext-Aware Recommendation Method

The proposed recommendation method is conceptually comprised of three main steps. The first step is to extract the candidate recommendation information from recommendation rules related to the context values. Following this, the recommendation information in the near future is extracted using the current context value and recommendation information rules. The detailed method is shown in [4].

In the second step, the recommendation information appearing in the very near future is pre-fetched dynamically using a context value, preferences and behavior for each context. The recommendation information extracted in the first step can cause the storage capacity and transmitting rate to increase. To solve these problems, only the recommendation information that can be used in the very near future is stored, and this process is then repeated.

In the ubiquitous application, there are a significant number of contexts vying for a user's attention [5]. Attention is defined as focused mental engagement on a particular message or piece of information [2]. The degree of attention may be determined using preferences and behavior; however, previous approaches have only used preferences. In this paper, attention is used as a weighting factor in the dynamic extraction process of a recommendation. The amount of recommendation information needed for a context with a high weight is larger than that needed for context with a low weight.

Our weighting mechanism works as follows: N is the number of context type, and C_n denotes context n. PW_n denotes the preferences weight, BW_n denotes the behavior weight and AW_n denotes the attention weight with respect to C_n. It is possible to compute AW_n using the following formula with a single adjustable parameter α ($0 \leq \alpha \leq 1$):

$$AW_n = (\alpha * BW_n) + ((1 - \alpha) * PW_n), \quad \text{where } 0 \leq BW_n, PW_n, AW_n \leq 1$$

$$PW_n = \frac{\text{the total number of } C_n \text{ in consumer's shopping records}}{\text{the total number of consumer's shopping records}}$$

$$BW_n = \frac{\text{the degree of consumer's behavior in } C_n}{\text{the maximum degree of consumer's behavior in } C_n}$$

The degree of a user's behavior is determined by a behavior policy. In the second step, the minimum degree of user's behavior is used in the BW_n, because the degree of user's behavior in the near future is not estimated. The BW_n is tuned using current degree of user's behavior in final step.

Moreover, the center and the periphery are adopted in our recommendation method. This approach engages both the center and the periphery of a user's attention, and in fact moves back and forth between the two. The periphery denotes what people are attuned to without attending to explicitly [6]. In this paper, while the user is in the center, all information extracted by attention weight is recommended, but when the user is in the periphery, only information extracted by attention weight above the threshold is recommended.

Our dynamic prefetching mechanism works as follows: the recommendation information that will appear in the very near future is prefetched from results of the first step using the context values and attention weight. The total amount of recommendation information that a user can obtain is restricted to a limited amount of information, called R_{max}. This means that the amount of the recommendation information of C_n does not exceed R_{max}. Thus, the amount of recommendation information to be prefetched is set to R_{max} multiplied by AW_n.

When a user's mobile device storage does not have enough space to store new recommendation results generated from the above step, the proposed method uses the MRU (Most Recently Used) replacement policy. Reference types of cache are categorized into three types: sequential, looping, and other references. In this paper, it is assumed that the reference type is sequential; meaning that consecutive block references occur only once. Sequentially-referenced blocks are never re-referenced; hence, the referenced blocks need not be retained. Therefore the MRU replacement policy is used [7].

In the final step, the recommendation information is composed dynamically. As BW_n is changed according to current degree of a user's behavior and the total amount of the results from the second step can exceed R_{max}.

Because BW_n is changed according to current degree of a user's behavior in the final step, the AW_n is needed to recalculate. If the current AW_n is larger than the AW_n resulting from the second step, the information is additionally delivered from the candidate recommendation information.

When the total amount of recommendation information from the second step exceeds R_{max}, the recommendation information for each context is rescaled into the range $[0, R_{max}]$ by min-max normalization.

4 Conclusion

Ubiquitous applications adhering to the anytime, anywhere, anydevice paradigm are required to be customizable meaning the adaptation of their services towards a certain context. Context-aware recommender systems are a core component in ubiquitous computing. The total amount of information is increased due to the greater number of contexts, and this requires an accurate and rapid recommendation method. However, previous recommendation approaches generally support only a single context. A new multicontext-aware recommendation method is suggested that uses attention composed of user's preferences and behavior. The composition of the most significant information for each context was changed dynamically over time such that it accurately reflected the user's attention. Our proposed method using user's attention is capable of recommending accurately and rapidly in ubiquitous computing.

References

1. P. J. Brown and G. J. F. Jones, "Context-aware Retrieval: Exploring a New Environment for Information Retrieval and Information Filtering", Personal and Ubiquitous Computing, Vol. 5, Issue 4, p.253-263, Dec., 2001.
2. T. H. Davenport, "May We Have Your Attention, Please?", Ubiquity Vol. 2, Issue 17, 2001.
3. G. Kappel, B. Proll, W. Retschitzegger and W. Schwinger, "Customisation for Ubiquitous Web Applications - A Comparison of Approaches", International Journal of Web Engineering and Technology, Vol.1, No.1, p.79-111, 2003.
4. J. Kwon, S. Kim and Y. Yoon, "Just-In-Time Recommendation using Multi-Agents for Context-Awareness in Ubiquitous Computing Environment", Lecture Notes in Computer Science 2973, p.656-669, Mar, 2004.
5. P. Tarasewich, "Designing Mobile Commerce Applications", Communications of the ACM, Vol. 46 , Issue 12 , p.57- 60, 2003.
6. M. Weiser and J. S. Brown, "The Coming Age of Calm Technology", Xerox PARC, Oct. 5, 1996.
7. J. Yoon, S. L. Min and Y. Cho, "Buffer Cache Management: Predicting the Future from the Past", Proceedings of the International Symposium on Parallel Architectures, Algorithms and Networks, p.105-110, 2002.

Context-Aware Trust Domains*

Ricardo Neisse**, Maarten Wegdam, and Marten van Sinderen

CTIT, University of Twente, The Netherlands
{R.Neisse, M.Wegdam, M.J.vanSinderen}@utwente.nl

Abstract. Context-aware service platforms need to establish and manage trust relationships for users to know if the user's privacy policies are being enforced and for service providers to control access to their services. Current trust solutions are not suitable for this because they do not address in an integrated manner trust issues related to identity provisioning, privacy enforcement and context information trustworthiness. In addition, due to their hierarchical and centralized design, current trust management solutions do not scale well in the ad-hoc pervasive environments in which context-aware platforms are typically deployed. In this paper we propose context-aware trust domains as a management solution for context-aware service platforms.

1 Introduction

One challenging problem in the realization of context-aware services [1] is the enforcement of the privacy of the users. This problem arises due to the highly privacy sensitive nature of the context information, and the implicit gathering and combining of this information in a pervasive service provisioning environment. For example, context information can be misused to allow unauthorized user tracking, unauthorized sophisticated user profiling and identity theft. It is therefore important for users to know about the trustworthiness [2] of the entities they are interacting with.

Next to being an object of security concern, context information also offers an opportunity to enhance the available security techniques. These enhancements include less intrusive access control methods where user roles are determined by context-situations (e.g. allow access for people inside a train [3]), instead of being assigned to specific entities. However, the use of context-information in this way requires trust (confidence) in the context-source, or requires at least a way to verify the integrity of the context information used in the access control policy (e.g. location).

In traditional systems, users establish static trust relationships with well known organizations such as banks, credit card companies, and mobile phone operators. These trust relationships are typically based on contracts, and the security policies are always associated with the entities' identities. For instance, a customer opening an account in a bank provides his/her personal data and, by signing a contract with the bank, establishes a trust relationship that his/her money and information will be stored

---

* This work is part of the Freeband AWARENESS project (http://awareness.freeband.nl). Freeband is sponsored by the Dutch government under contract BSIK 03025.
** Ph.D. student supported by CNPq scholarship – Brazil.

P. Havinga et al. (Eds.): EUROSSC 2006, LNCS 4272, pp. 234–237, 2006.

safely. In pervasive environments, users are supposed to interact with a large number of entities unknown beforehand and a priori trust relationships only exist in a few special scenarios where nodes are controlled by a single organization [4].

2 Trust Aspects

In order to scope our work we divide Trust in different aspects as [5] approaches Privacy. Trust can be analyzed regarding the social, the informational, and the technical aspects (Fig. 1). For each of these aspects there are different problems that should be addressed, for instance, how user perceive the trust in the system (social aspect), what are the concepts and semantics of trust mapped into the system (informational aspects) and how secure is the encryption technology used (technical aspect). The main focus of this work is in an architecture for dynamic management of trust relationships (information aspect) but we are also interested in the social aspect.

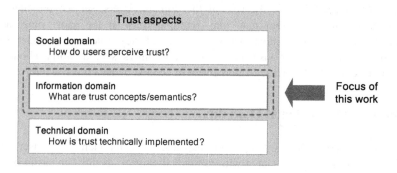

Fig. 1. Trust aspects

3 Trust Relationships in a Context-Aware Service Platform

Fig. 2 presents a context-aware service platform [6] on which we base our proposal for context-aware trust domains. In this service platform, before accessing a context-aware service, users should authenticate (step 1) with an identity provider. After the authentication is done, users can start using the service (step 2), which may check the user identity (step 3). The context-aware services retrieve context information about the users from context sources (step 4) in order to provide a personalized response. This context-information can be of any type, for instance, information representing the current activity or location of the user.

The four main roles we have identified in this context-aware service scenario (Fig. 2) are: *identity providers, context consumers* (service providers), *context providers* (context sources) and *context owners* (e.g. users). In Fig. 2 we also present trust relationships from users (a and b) and service providers (c) points of view. The list of trust relationships presented is not exhaustive; other different trust relationships could be defined. In order to limit the scope we focus on trust relationships related to privacy (i), identity (ii), and context trustworthiness (iii).

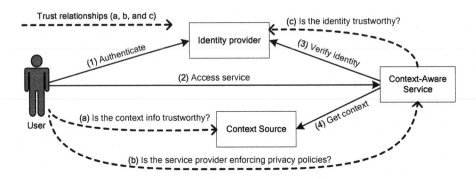

Fig. 2. Trust relationships in a context-aware service platform

We consider that each entity in the context-aware service platform is part of an administrative domain for which a common set of trust relationships and security policies applies, according to the role of the domain. Users, service providers and context-sources in the system may play multiple roles at the same time, acting as context-consumers and context-owners. For each of these roles, we consider different trust requirements, for instance, context consumers should worry about trust relationships related to the privacy enforcement of context owners (user's) information.

3   Context-Aware Trust Domains

In a large context-aware system, with thousand of components and users, trust relationships can not be associated with individual entities, as this can easily become unmanageable. The main objective of this work is to reduce the complexity in the management of trust relationships using the abstraction of context-aware domains. In this way, trust degrees do not have to be specified individually for each entity, but in a set for a collection of entities part of a domain [7].

Domains are useful in the management because they allow the association of trust and policies with collections of entities instead of individual entities. In Context-aware domains, context information is used as a dynamic constituent, allowing more flexibility in the domain definitions. In this way, entities sharing the same context are grouped together and can join/leave the domain dynamically, with the context changes. Examples of context-aware management domain definitions are "Nearby persons", "Personal devices", and "Working colleagues" (Fig. 3).

Our proposal is to elaborate on the concept of Context-Aware domains for trust and policy management, where entities inside of a specific context situation will be associated with a trust degree and/or security policy. The idea is to provide mechanisms to define and infer the trust degree of an entity based on the context information provided about that entity. We divide this research in two main open research questions: (1) what is the role of trust in context-aware service platforms and (2) what is the role of context information in trust establishment and management.

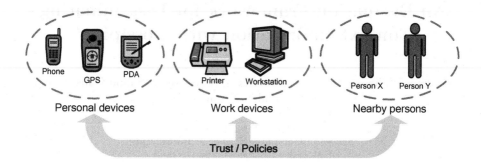

Fig. 3. Examples of context-aware domains

We plan to validate our context-aware trust domain concept through a prototype implementation in the AWARENESS project. The validation method probably will be focused on the performance and usability and based on analysis of system logs/traces and end users and system administrators interviews.

References

1. Dey, A. K.; Salber, D.; Abowd, G. D. A conceptual framework and a toolkit for supporting the rapid prototyping of context-aware applications. HC Interaction, 16, 2001.
2. Grandinson, T.; Sloman, M. A Survey of Trust in Internet Applications, IEEE Communications Surveys 2000.
3. Hulsebosch, R. J.; Salden, A. H.; Bargh, M. S.; Ebben, P. W.; Reitsma, J. Context sensitive access control. In Proceedings of the 10th ACM SACMAT, Sweden, June, 2005.
4. Molva, R. and Michiardi, P. Security in Ad hoc Networks (invited paper). In: Personal Wireless Communications, September 23-25 2003, Venice Italy.
5. Boland, H.; Soute, I. *Perceived Privacy*. Philips research presentation, AMIGO project workshop, Telematica Instituut, The Netherlands, March, 2006.
6. AWARENESS Service Infrastructure D2.1 - Architectural specification of the service infrastructure, http://awareness.freeband.nl.
7. Damianou, N.;Dulay, N.; Lupu, E.; Sloman, M.; Tonouchi, T. Tools for Domain-based Policy Management of Distributed Systems. IEEE/IFIP NOMS, Italy, Apr, 2002.

An Evaluation Framework for Disseminating Context Information with Gossiping*

Graham Williamson, Graeme Stevenson, Steve Neely, Simon Dobson, and
Paddy Nixon

Systems Research Group
School of Computer Science & Informatics
University College Dublin, Ireland
graham.williamson@ucd.ie

Abstract. As we gain access to increasing volumes of context data,
we face the problem of moving this information from the sensors that
produce it to the applications that consume it. Our approach to this
problem uses *gossiping*, a probabilistic routing protocol, to disseminate
context information throughout the environment. We present on-going
work on evaluating the performance of different gossiping protocols for
this purpose.

1 Introduction

Pervasive computing systems require a large amount of information to be available in order to support adaptive, context-aware, applications. We face the challenge of delivering information from contributing sensors to all points where it is
required by applications. Furthermore, because of the nature of the information
and the applications which use it, it must be delivered in a timely manner.

The characteristics of data and applications in these systems allow us to use
a non-deterministic mechanism called *gossiping* [1] to underpin the communications. In order to explore the performance of gossiping for disseminating context
information, after a brief overview of gossiping, we present ongoing work on an
evaluation framework for gossiping protocols.

2 Gossiping in Pervasive Systems

Data from sensors in these systems, in addition to having a limited lifetime, are
often frequently repeated. For example, the location of a person will be continuously updated; or the reading from a temperature sensor may be refreshed
periodically. The implication of this is that we do *not* need to guarantee complete
reliability of message delivery to all nodes: a missed message is likely to be refreshed. Relaxing the reliability constraints allows us to use *gossiping* to support

---

* This material is based on works supported by Science Foundation Ireland under
Grant No. 04/RP1/I544.

P. Havinga et al. (Eds.): EUROSSC 2006, LNCS 4272, pp. 238–239, 2006.

communications. We propose that gossiping can provide the desirable properties of scalability, decentralisation, and robustness to change, that are required for pervasive computing systems.

Gossiping is decentralised and uses only simple, local interactions. This produces emergent behaviour which can result in a scalable, resilient network. However, because behaviour is emergent, it can be difficult to evaluate the effectiveness of a specific algorithm, or the effect that can be produced by small tweaks in the same algorithm.

3   The Evaluation Framework

The gossiping evaluation framework is a two-phase approach involving simulation in OMNet++ [2], a discrete event simulator that can model computer networks and communication protocols, and real-world experimentation on Planet Lab [3]. These methods compliment and provide validation against each other, allowing us to come to a consensus on the performance of a particular implementation.

We identify the set of parameters that can affect the performance of a gossiping implementation (such as the frequency of gossips, or the number of nodes gossiped to on each round of the algorithm). The impact of altering each parameter must be assessed in order to gauge performance. Thus, we also identify a set of measurements that we feel can be used to characterise and compare the performance of gossiping algorithms.

The key properties that we will assess initially using our evaluation framework inlcude the:

- **latency** of data propogation through networks of different topologies
- **coverage** of message dissemination with respect to time
- **robustness** of the data propogation under node and link failures
- **scalability** of the system with respect to the number of participating nodes.

4   Conclusion

With this evaluation framework we will confirm the suitability of gossiping for the dissemination of context information and we will examine the effect that a number of parameters can have on the performance of gossip style algorithms. The results from these experiments will allow us to tune our implementation and gain an understanding of the issues and trade-offs involved when designing a gossiping protocol.

References

1. Eugster, P.T., Guerraoui, R., Kermarrec, A.M., Massoulie, L.: Epidemic information dissemination in distributed systems. Computer **37** (2004) 60–67
2. Varga, A.: The OMNet++ discrete event simulation system. In: Proceedings of the European Simulation Multiconference (ESM'2001), Prague, Czech Republic (2001)
3. The PlanetLab Homepage. (http://www.planet-lab.org/) Last accessed: July, 2006.

Dynamic Bayesian Networks for Visual Surveillance with Distributed Cameras

Wojciech Zajdel[1], A. Taylan Cemgil[2], and Ben J.A. Kröse[1]

[1] Informatics Institute, University of Amsterdam
Kruislaan 403 1098SJ Amsterdam
{wzajdel, krose}@science.uva.nl
[2] Signal Processing and Communications Laboratory
University of Cambridge
atc27@cam.ac.uk

Abstract. This paper presents a surveillance system for tracking multiple people through a wide area with sparsely distributed cameras. The computational core of the system is an adaptive probabilistic model for reasoning about peoples' appearances, locations and identities. The system consists of two processing levels. At the low-level, individual persons are detected in the video frames and tracked at a single camera. At the high-level, a probabilistic framework is applied for estimation of identities and camera-to-camera trajectories of people. The system is validated in a real-world office environment with seven color cameras.

1 Introduction

Societal developments, like the aging of population, globalization, increased mobility have lead to an increased demand for computer systems for assistance in safety, comfort or communication. These trends, together with steady advances in technology, inspire research on "smart" environments that observe the users and make decisions on the basis of the measurements. Examples are safety systems for elderly, traffic control systems, or security systems in public places. In order for the systems to be "aware" of what is going on, these systems are equipped with various sensors, such as infrared detectors, radar, cameras, microphones, mounted at many locations.

As a part of this research field, we developed a surveillance system for tracking multiple people through a wide area with sparsely distributed cameras. The core of the system is an adaptive probabilistic model for reasoning about appearances, locations and identities of people. Our system consists of two processing levels. At the low-level, individual persons are detected in the video frames and tracked at a single camera. At the high-level, a probabilistic framework is applied for estimation of identities and camera-to-camera tracking of people.

2 Probabilistic Model for Multi-camera Tracking

The task of the high level system is to maintain person's identity when he or she leaves the filed of view of one camera and later on appears at some other camera.

P. Havinga et al. (Eds.): EUROSSC 2006, LNCS 4272, pp. 240–243, 2006.

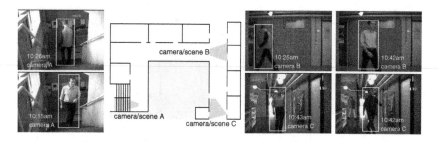

Fig. 1. An example of the considered tracking problem, with three cameras: 'A', 'B' and 'C' that observe non-overlapping scenes. Every image depicts a complete pass of a person through a camera viewing field.

Since the cameras in our system are sparsely distributed, their fields of view are disjoint. Consequently, when a person leaves the field of view we temporarily loose track (see Fig. 1). When the person appears again at some other camera we aim to re-identify the person on the basis of appearance and global motion constraints (like the minimum travel time between the two camera locations).

For this purpose we developed a probabilistic model where identities are represented as discrete random variables called labels. The model assumes that the labels are latent (hidden) and have to be estimated from of measurements of tracked objects (people). A single measurement represents a complete pass of a single person through a camera field of view. The measurements are provided by the low-level system and consist of two components: appearance features (various color statistics) and spatio-temporal features (location, time).

The dependency between the labels and measurements is expressed as a probability density in the form of directed graphical model, as illustrated by Fig. 2. Aside of labels (s_k) and measurements (y_k), the model includes latent state variables (x_k) and auxiliary latent pointer variables (z_k), where $k = 1, \ldots$ denotes the measurement index. The (continuous) state variable represents intrinsic color properties of a person. The auxiliary variable z_k is a collection of (discrete) values that indicate the number of persons up to measurement k and, for every object, indicate the previous measurement of that object (see [6] for details).

We consider the appearance features of y_k as generated by a Gaussian density parametrized by the state x_k (mean, covariance matrix). If the label s_k indicates a person not observed before, then the state (i.e., the parameters of Gaussian kernel) is generated from a prior state density. Otherwise, the state x_k is set equal to the state that generated the previous measurement with the label s_k. Formally, the model can be viewed as an instance of Infinite Gaussian Mixture Model (also known as Dirichlet Process Mixture Models) [4].

Given the model, we developed an efficient inference algorithm that estimates labels from the measurements. The algorithm computes marginal posterior densities in the form $p(s_k|y_{1:k})$ from which we can find the MAP label for every measurement. The resulting tracking method works in on-line regime, where a label s_k is estimated from the currently available measurements $y_{1:k}$. Importantly, the

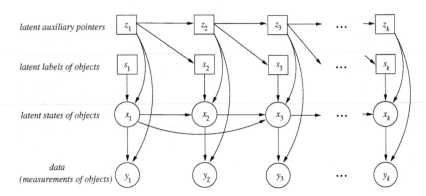

Fig. 2. The Dynamic Bayesian Network that underlies the wide-area tracking of people. Ovals indicate continuous domain variables, rectangles — discrete domain variables.

sought density $p(s_k|y_{1:k})$ cannot be computed exactly, since our model belongs to a class of hybrid graphical models where inference is intractable [1]. For approximate inference we applied an efficient technique known as the assumed-density filtering (ADF) [2].

3   Experiments and Results

We set up an experimental system and collected a dataset that includes 70 observations representing 5 persons who were observed in an office building with 7 cameras with disjoint views (see Fig. 3). To test our algorithm in various conditions we manipulated the constraints on camera-to-camera motion of people, and expressed the constraint strength as entropy value. Low-entropy (strong) constrains imply that people move along fixed paths (easier to track). We applied our algorithm and compared it with Multiple Hypothesis Tracking (MHT) and

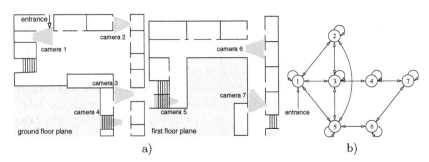

Fig. 3. (a) Building plan were the observations were taken. The gray areas show camera viewing fields (scenes). (b) A graph showing movement constraints assumed by distributions on the transitions. The numbered nodes indicate scenes, the edges indicate allowed scene-to-scene transitions.

with an MCMC approach as used by Russell in a highway context [3]. We also investigated how our method performs if only spatio-temporal features from the observations are used, and if only visual features from the observations are used. The basic results are given in Tab. 1, more experiments are presented in [5].

Table 1. Tracking in environments with varying difficulty level

entropy	accuracy [%] ADF	MHT	MCMC	recall [%] ADF	MHT	MCMC	objects ADF	MHT	MCMC
1.17^a	96	77	90	94	59	85	5	8	7
1.21^a	65	81	90	83	33	84	4	18	7
1.43^a	72	83	91	83	30	78	5	21	8
2.23^a	66	79	90	79	24	60	4	21	13
2.77^b	61	71	77	60	20	46	5	24	13
1.17^c	59	71	70	63	68	70	5	5	6

[a] Tracking based on spatio-temporal and appearance features.
[b] Tracking based exclusively on appearance features.
[c] Tracking based exclusively on spatio-temporal features.

4 Conclusions

Results show that our methods are much more robust to illumination conditions, robust to specific shapes, have a better performance in tracking and scale much better with the number of cameras and the number of humans in the system. As a result, parts of the presented techniques are currently being tested by industrial partners for potential applicability in their surveillance systems.

References

1. U. Lerner and R. Parr. Inference in hybrid networks: Theoretical limits and practical algorithms. In *Uncertainty in Artificial Intelligence*, pages 310–318, 2001.
2. K. Murphy. *Dynamic Bayesian Networks: Representation, Inference and Learning.* PhD thesis, University of California, Berkeley, 2002.
3. H. Pasula, S. Russell, M. Ostland, and Y. Ritov. Tracking many objects with many sensors. In *Int. Joint Conf. on Artificial Intelligence*, pages 1160–1171, 1999.
4. C. E. Rasmussen. The infinite Gaussian mixture model. In *Advances in Neural Information Processing Systems 12*, pages 554–560, 2000.
5. W. Zajdel. *Bayesian Visual Surveillance. From object detection to distributed cameras.* PhD thesis, University of Amsterdam, 2006.
6. W. Zajdel, N. Vlassis, and B. J. A Kröse. Bayesian methods for tracking and localization. In E. Aarts, J. Korts, and W. Verhaegh, editors, *Intelligent Algorithms*, pages 243–258. Kluwer Academic Publishers, 2005.

Embedded Intelligence: Enabling In-Situ Power Management for Wireless Sensor Networks

Rui Ma[1], Gregory M.P. O'Hare[2], and Michael J. O'Grady[2]

[1] School of Software, Beijing Institute of Technology
5 South Zhongguancun Street, Haidian District, Beijing 100081, P.R. China
mary@bit.edu.cn
[2] Adaptive Information Cluster, School of Computer Science and Informatics
University College Dublin, Belfield, Dublin 4, Ireland
{gregory.ohare, michael.j.ogrady}@ucd.ie

Abstract. Effective and efficient power management remains one of the most formidable obstacles that must be overcome before Wireless Sensor Networks can be deployed on a widespread basis. Embedding agents on individual sensors offers one promising approach for intelligent power management. In this paper, initial results of simulations are presented which indicate the potential of this approach for extending the lifetime of individual sensor nodes, and ultimately, Wireless Sensor Networks.

1 Introduction

Wireless Sensor Networks (WSNs) have been successfully deployed in a wide variety of application domains but, while their potential is widely acknowledged, such networks remain compromised in one crucial aspect: power. As the lifespan of the network is, in essence, determined by the operating lifetime of its constituent nodes, effective power management becomes an indispensable issue in the design and deployment of WSNs [1]. Indeed, resolving this issue is essential if pervasive computing and the widely-anticipated Ambient Intelligent (AmI) [2] vision is to become a reality.

In a standard sensor node, the transceiver consumes the most power, suggesting that transmission occurrences should be minimised. However, reconciling the conflicting demands of maximising the lifetime of individual sensor nodes while simultaneously maintaining the quality of information flowing from the WSN is a non-trivial task; and may be heavily influenced by the characteristics of the particular application domain in question. Intelligent techniques, and in particular, intelligent agents, offer an intuitive mechanism for modelling and resolving some of these conflicts.

This paper investigates the efficacy of using intelligent agents, and in particular, Belief-Desire-Intention (BDI) agents [3], in delivering in-situ power management for wireless sensor nodes.

P. Havinga et al. (Eds.): EUROSSC 2006, LNCS 4272, pp. 244–247, 2006.

2   Simulation Architecture

To verify the feasibility of an agent-based approach, it was necessary to integrate two software tools, namely J-Sim [4] and Agent Factory Micro Edition (AFME) [5].

- J-Sim is an open-source, component-based, compositional simulation environment that provides a WSN simulation environment. Power models (both consuming and producing) are included.
- AFME constitutes a framework which supports the deployment of intelligent agents on devices that are computationally resource poor, for example, PDAs and mobile phones.

We undertook simulations within J-Sim where AFME agents were deployed on the node and infused with two rudimentary rules for reducing energy consumption.

1. The agent continuously monitors the battery level and when this is below a certain threshold, the operational mode of the radio is changed. The radio may be in one of five states (idle, sleep, off, transmit and receive). In certain situations, changing the radio state from, for example, receiving to sleep state, may conserve energy.

$$\text{BELIEF(energy(?val))} => \text{COMMIT(Self, Now, BELIEF(true),} \qquad (1)$$
$$\text{ChangeRadioMode(?val))}$$

2. Reducing the sampling rate will reduce the energy being consumed on the sensor.

$$\text{BELIEF(samplingRate(?rate)) \& BELIEF(energy(?val))} => \qquad (2)$$
$$\text{COMMIT(Self, Now, BELIEF(true), AdoptSamplingRate(?rate, ?val))}$$

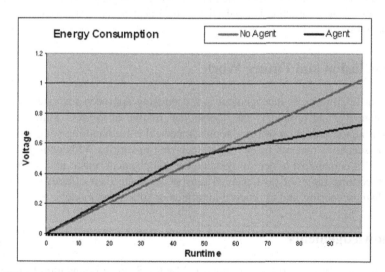

Fig. 1. Once a certain voltage threshold is surpassed, the radio mode is adapted, reducing the rate at which energy is consumed

The results of the simulations are shown in Figures 1 and 2 respectively. In Figure 1, although the sensor node will consume more energy when the agent is embedded within it, the node's lifetime will be extended because the agent will opportunistically adjust the radio mode between the receiving, transmitting and sleeping modes. In Figure 2, when we set the sampling rate, the sensor node consumes less energy as distinct from the case where the receiving signal is being monitored at all times.

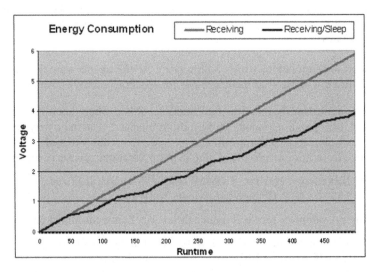

Fig. 2. At certain threshold values of the available energy and current sampling rate, a rule is activated by the agent resulting in a reduction of the sampling rate and a reduction in energy consumption

3 Conclusion and Future Work

Intelligent in-situ power management is a promising approach for extending sensor node longevity, and ultimately the operational lifetime of WSNs. Future work will focus on the identification of more sophisticated rules for handling some of the more common scenarios that can be envisaged recurring in various WSN configurations. In particular, it is intended to deploy agents on actual physical sensor nodes, specifically the forthcoming range of java-enabled sensors, evaluate their effectiveness in small but realistic scenarios, and to proceed to refine our simulation models.

Acknowledgements

This material is based upon works supported by the Science Foundation Ireland (SFI) under Grant No. 03/IN.3/1361.

References

1. Hempstead, M., Tripathi, N., Wei, G. Y., Brooks, D., An Ultra Low Power System Architecture for Sensor Network Applications. Proceedings of the 32nd Annual International Symposium on Computer Architecture. IEEE Computer Society, Washington, DC, USA (2005) 208-219
2. Vasilakos, A., Pedrycz, W., Ambient Intelligence, Wireless Networking, Ubiquitous Computing, Artec House, 2006.
3. Rao, A.S., Georgeff, M.P., Modelling Rational Agents within a BDI Architecture. In: Principles of Knowledge Representation. & Reasoning, San Mateo, CA. 1991.
4. Sobeih A., Chen W. P., Hou J. C., Kung L. C., Li N., Lim H, Tyan H. Y., Zhang H. H.: J-Sim: A Simulation Environment for Wireless Sensor Networks. Proc. of the 38th Annual Symposium on Simulation, Washington DC, USA (2005) 175-187. IEEE
5. Muldoon, C., O'Hare, G. M. P., Collier, R. W., O'Grady, M. J.: Agent Factory Micro Edition: A Framework for Ambient Applications. Proceedings of the Intelligent Agents in Computing System Workshop, International Conference on Computational Science. Reading, UK, 2006.

Proximity Sensing Using IEEE 802.15.4 Radios

Mark Lowton, James Brown, Joe Finney, and Gerd Kortuem

Computing Department
Lancaster University
Lancaster, LA1 4WA, UK
{lowton, jb, joe, kortuem}@comp.lancs.ac.uk

Abstract. Accurately determining the location of devices is a key challenge in contextual smart sensing. This poster describes how IEEE 802.15.4 radios can be used for accurate proximity sensing of co-located devices. In particular, the effect of low noise amplifiers on the estimation accuracy are discussed.

1 Introduction

The NEMO research project [1], at Lancaster University, is investigating using contextual smart sensing to improve health and safety (H&S) at industrial sites. Work activities at such sites are governed by H&S regulations designed to minimise risks to workers. The technical approach of NEMO is to augment, in the workplace, artefacts such as tools, vehicles and workers, with smart devices to collect and contextually analyse data from their environment to detect (and potentially correct), violations in H&S regulation in the field [2]. Consider, for example, a condition known as vibration white finger (VWF) which incurs the (total) loss of feeling in the fingers. Since its discovery, H&S rules have been introduced to limit a worker's exposure to vibration. By augmenting tools with smart sensing devices, vibrations can be measured and recorded in the field and keep a worker's exposure within the regulations.

Many H&S regulations rely on the concept of close quarters proximity. For the VWF example, proximity between a tool and a worker offers strong hints to which tool a worker is operating. There are also, for example, regulations which govern the storage of materials in chemical plants, requiring knowledge of the relative position or proximity between containers [2]. To enable these applications, there is a need to accurately, reliably, and inexpensively detect the distance between devices but only within a small, fixed distance of one another – *typically a few metres.*

Whilst proximity can be measured using custom solutions (perhaps RFID), it is beneficial to minimise complexity and promote ease of deployment and energy efficiency. These requirements rule out the use of GPS or Wi-Fi based solutions. The poster shows how, in practice, proximity can be inferred from the RF characteristics of modern data radios, in particular IEEE 802.15.4 [5] compliant radios.

2 Proximity Measurement from RF Channel

Previous investigations attempting to marry the strength of a received radio signal and locality have offered useful findings [3][4]. It is understood the impact of the environment on radio wave propagation offers unclear values in the received signal

P. Havinga et al. (Eds.): EUROSSC 2006, LNCS 4272, pp. 248–249, 2006.
© Springer-Verlag Berlin Heidelberg 2006

strength (RSS), presenting indistinguishable trends as a receiver migrates about a transmitter. It has been seen that (in a static environment) such results are consistent, but poor for determining distances [6].

Advanced radio receivers possess variable gain low noise amplifiers (LNAs) which reactively amplify input signals to a more 'workable' level. This is typically performed before any other on the radio, *including RSSI detection*, providing benefits for data reception, but adding error into the perceived RSS. To understand the impact of LNA on effectiveness of RSSI based techniques, its effects were investigated using IEEE 802.15.4 compliant radio transceiver modules. The investigations indicated that the RSSI anomalies usually observed are the direct result of using LNA and show distance estimations are more robust for a fixed LNA setting.

3 Findings

It was found that by attenuating the output of an IEEE 802.15.4 transmitter, the transmission range can be effectively reduced; this maintains the resolution but increases the granularity of RSSI levels. In addition, fixing the LNA gain offers increased stability in RSSI over a short transmission range (1-2m).

Implementing these findings revealed that close quarters proximity detection can be achieved using this technology, thus enabling applications such as the health and safety scenarios, without the need for additional equipment on the embedded device.

4 Conclusion

For close quarters proximity (1-2m), RSSI may be used as a metric to infer locality and that this is possible to an accuracy of at least 10cm. This can be achieved by attenuating a transmitter and limiting the LNA of a receiver radio, to gain a usable correlation between distance and RSSI over a short distance.

The experiments reported in this poster can be found in full in [6].

References

1. NEMO project web site www.comp.lancs.ac.uk/nemo
2. Strohbach, M., Gellersen, H., Kortuem, G., Kray, C. Cooperative Artefacts: Assessing Real World Situations with Embedded Technology, Ubicomp '04, Springer, Berlin, Heidelberg, N.Y. pp 250-267
3. Hightower, J., Boriello, G., Want, R., SpotON: An Indoor 3D Location Sensing Technology Based on RF Signal Strength, UW CSE 2000-02-02, University of Washington, Feb 2000
4. Bahl, P., Padmanabhan, V.N., RADAR: An In-Building RF-based User Location and Tracking System, INFOCOM 2000. 19th Conference of IEEE Computer and Communications Societies pp.775-784 vol.2
5. IEEE 802.15.4 Standard: Wireless Medium Access Control and Physical Layer Specification for Low Rate Wireless Personal Area Networks (LR-WPAN), May 2003. Version 802.15.4-2203
6. Lowton, M., Brown J., Finney, J., Finding NEMO: On the Accuracy of Inferring Location in IEEE 802.15.4 Networks, REALWSN '06. Workshop on Real-World Wireless Sensor Networks, June 2006

Towards Hovering Information

Alfredo Villalba and Dimitri Konstantas

Centre Universitaire d'Informatique, University of Geneva,
24 rue Général Dufour, 1211 Geneva 4, Switzerland
{alfredo.villalba, dimitri.konstantas}@cui.unige.ch

Abstract. This paper introduces a new concept of information that can exist in a mobile environment with no fixed infrastructure and centralized servers, which we call the Hovering Information. This information will be capable of staying attached to a specific geographical point, hovering from one device to another for surviving or even moving from one place to another as defined by its creator or other factors. Its applications can be many. In this paper we describe two main scenarios: disaster areas and tagged world, discussing related issues such as persistency and reliability, distribution, consistency and security.

1   Concept

In the future, we can safely assume that daily life objects and people will be equipped with a mobile device having large memory capacity, computing power and connected to a wireless highly speed network. Eventually the world will become a fully interconnected network where each person and object will be a node capable of creating, inserting, accessing, storing, processing or diffusing a new type of information. This information will no longer be stored in large fixed servers, but it will be available everywhere in the environment. Eventually this information will detach itself from the physical media and will have the ability of staying attached to a location (in space and time) hovering from one storage device to another, or even moving from one place to another, as possibly specified by its creator, or defined by the information itself. This new type of information that we call *"hovering information"*, will not be linked to any storage device and will have an existence of its own, linked to a location (rather than a hardware device) that we call the *anchoring location*.

The recent progress in mobile networks, mobile ad-hoc networks, sensor networks, RFIDs and other related technologies have already given a step towards this vision. Thanks to mobile wireless networks, it is today possible for two or more persons to communicate and exchange multimedia messages using just their mobile phone or PDA from almost anywhere on the globe. Combined with localization systems in mobile devices, like GPS and Galileo, sensors and RFIDs, sophisticated location based and context-aware services are becoming available to mobile users.

The concept of hovering information aims in bringing together all these technological advances providing the concepts and eventually a platform that will allow the development of user-oriented services and applications.

P. Havinga et al. (Eds.): EUROSSC 2006, LNCS 4272, pp. 250–254, 2006.

2 Applications

The applications of hovering information are limited only by our imagination. Nevertheless, in order to demonstrate the basic concepts and ideas we describe two representative scenarios that will clarify the nature and behavior of the hovering information and its potential usage.

2.1 Disaster Areas

It is very probable that after a natural disaster (e.g. earthquake, tsunami, etc.) the communications' infrastructure will be damaged and out of operation. As a result in the emergency situation where the persons on location will require urgently information, nothing will be available. However it will be also very probable that many stand-alone mobile or fixed devices (e.g. RFIDs attached to objects or mobile phones) will be still operational and capable of establishing an ad-hoc communication network. It is exactly in this case where hovering information can provide the required means for the dissemination of needed information. Since people will be continuously moving from one location to another, trying to help someone or finding a missing person, or even marking dangerous areas, information will need to be linked to locations instead of devices (that will be moving continuously). Disaster survivors can for example post information regarding places that need to be searched, names of missing people, etc, while rescue workers can mark dangerous areas, provide rescue instructions and directions, etc.

People will try to leave as soon as possible the disaster area. However the information they created will find its way (as hovering information) to stay in place by migrating from one available device to another. In this way people will be able to access and enrich the information without the need of fixed networked servers. As the situation in the disaster area improves, new information will appear, while the older one will disappear or migrate wherever it might be needed.

2.2 Tagged World

In the previous example the hovering information can be described as being ephemeral: it is bound to disappear after some time. However we can also imagine persistent hovering information. During the last few years several location based services have been proposed allowing users to place virtual tags at different places in space (called in the literature "air graffiti", "space tags" etc). All these services require a centralized server where the information (along with its space coordinates) is stored. The concept of the hovering information however offers and alternative way to link information in space, without the need of a central server. The advantages of using the hovering information concept instead of a centralized solution for the creation virtual tags, are many. First of all there is no central control. This is a major advantage from the point of view of service robustness, since no central control means no single point of failure.

A second advantage is that the hovering information allows far greater parallelism in accessing the information at a certain location, thus allowing eventually faster access to the information. A third advantage (which of course can become disadvantage) is that no-one can control, remove or censor the information. Hovering information provides full user empowerment for the distribution of information. Finally the use of hovering information can valorize the increasing and underused storage and processing capacity of mobile devices. There are many terabytes of storage capacity available in mobile devices that are heavily underused. Hovering information can provide the means to make a good use of this storage capacity.

3   Issues

The implementation of the hovering information concept has many implications ranging from social and cultural issues, to technical problems and design questions. We will describe only some of the research questions and technical issues, related to the implementation of the concept.

3.1   Persistency and Reliability

The persistency and reliability of hovering information is one of the most important problems to solve due to the mobile nature of the storage nodes and the absence of a fixed infrastructure. Hovering information stored at a location might disappear if all nodes present leave the location at some time. In this case, there will be no node left in the location or near by. When the last node will leave the anchoring location vicinity, the hovering information will disappear or will go away with the last node and will no longer be accessible at the specific location. However, the disappearing of information that is no more useful could be an advantage to be considered.

3.2   Consistency

The consistency of hovering information presents some problems since the information that was posted by a person, will be replicated or migrated, in order to ensure robustness, reliability, and persistency. As a result if the owner of the information wishes to update the information, it might be impossible to access all nodes where the hovering information is stored and update it. However we have to note that hovering information is like the knowledge exchange by humans. That is, every human has his version about some specific knowledge, which might or might not be the same as another person. The human race had managed to live long, and even take advantage, in spite the presence of this type of contradictory information! Thus we can imagine the existence of inconsistent hovering information may be in fact an advantage. That it might be useful to have many opinions about a subject and being able to choose what is the more convenient for us.

3.3   Distribution

The distribution of hovering information represents an interesting and may be a very important issue. Hovering information does not really needs to be anchored in a specific location, but it can move freely from one place to another. This nomadic hovering information will move from one place to another, as defined by its creator or by the information itself, for example, depending where the information is needed. We can even imagine hovering information migration, when many persons (nodes) request to access this information at a specific location. For example, we might observe the migration of football related information to Germany, during the world cup, taking place in Germany, where thousands of fans are requesting related information.

3.4   Security

The nature of hovering information is dynamic and user defined. These two factors are an advantage and at the same time a source of security challenges. Any person or any autonomous device will be capable of creating and inserting information about some subject and this information will be anchored at a location or distributed around the world. Since there is no central control and no predefined communication channels, we will need to develop new security mechanisms that will allow us to protect nodes from malicious information, chase and erase information that should not be there, provide some means to define trust in the information etc. This latest, that is trust, is one of the most interesting points to work on, due to its direct relation with the nature of hovering information. For instance, how can one trust some information saying that there is a souvenir shop near a place (which is not true) and this information has been posted by a thief for attracting victims to his location.

4   Related Work

The concept of Hovering Information being original, there is, in our knowledge, no directly related past work. We will thus make use of work coming from different areas as background for the development of our ideas. PeopleNet [1] describes a wireless virtual social network which mimics the way people seek information via social networking. It uses the infrastructure to propagate queries of a given type to users in specific geographic locations, called bazaars. Within each bazaar the query is further propagated between neighboring nodes via peer-to-peer connectivity until it finds a matching query. Serendipity [2] is a project from MIT Media Labs, it provides simple user interfaces on mobile phones where a user can post his/her profile and place a query. If two users are within Bluetooth range and share similar profiles they are immediately alerted. The queries propagate only through the mobile ad-hoc network. The work described in [3] is a system of collaborative backup. It relies on collaboration among peers in order to provide data backup and recovery, all this in a mobile environment. Finally, 7DS [4] is peer-to-peer resource sharing system enabling the exchange of data among peers that are not necessarily connected to the Internet, these peers can be either mobile or stationary.

5   Research Directions

One assumption we made in describing the hovering information concept and applications, is the existence of powerful mobile devices (with lots of memory, processing power and high bandwidth wireless communications) carried by people or attached to objects. Today however, mobile networks have a limited bandwidth, routing protocols over mobile ad-hoc networks are not yet sufficiently performing, sensor networks and RFIDs have a limited range communications, memory and computing power. This possess more problems and restrictions to the study and the design of a platform supporting the hovering information concept, but we are confident that in the near future the evolution of the technology will allows to implement the system in its full scale and capabilities.

The study of mobility patterns, route prediction algorithms and social networks' dynamics could give us some ideas about the trajectories of storage nodes and so how to design storage-replication-distribution algorithms in order to guaranty the persistency of the hovering information and its migration. Trust management may give us some starting points for studying the trust of hovering information, and as we said before, the existence of contradictory hovering information could be an advantage in some applications. Finally, mobile ad-hoc networks and peer-to-peer paradigms could give us some ideas concerning the architecture of a platform supporting hovering information.

Our next steps will be to refine the concept of the hovering information, defining the conceptual and technical challenges and providing some type of solutions. We will then proceed to a design and the development of a prototype for the study of related issues.

References

1. Motani, M., Srinivasan, V., Nuggehalli, P.S.: PeopletNet: Engineering a Wireless Virtual Social Network. Proceedings of the 11th Annual International Conference on Mobile Computing and Networking. ACM Press, New York, NY, USA (2005) 243–257
2. Eagle, N., Pentland, A.: Reality Mining: Sensing Complex Social Systems. Personal and Ubiquitous Computing, Vol. 10, No. 4. Springer-Verlag London Ltd (2006) 255–268
3. Courtès, L., Killijian, M.-O., Powell, D., Roy, M.: Sauvegarde Coopérative entre Pairs pour Dispositifs Mobiles. Proceedings of the 2nd French-speaking Conference on Mobility and Ubiquity Computing. ACM Press, New York, NY, USA (2005) 97–104
4. Papadopouli, M., Schulzrinne, H.: Effects of Power Conservation, Wireless Coverage and Cooperation on Data Dissemination among Mobile Devices. Proceedings of the 2nd ACM International Symposium on Mobile ad-hoc Networking and Computing. ACM Press, New York, NY, USA (2001) 117–127

Balancing Smartness and Privacy for the Ambient Intelligence

Harold van Heerde[1], Nicolas Anciaux[2], Ling Feng[1], and Peter M.G. Apers[1]

[1] Centre for Telematics and Information Technology
University of Twente, The Netherlands
{h.j.w.vanheerde, ling, apers}@ewi.utwente.nl
[2] INRIA, France
Nicolas.Anciaux@inria.fr

Abstract. Ambient Intelligence (AmI) will introduce large privacy risks. Stored context histories are vulnerable for unauthorized disclosure, thus unlimited storing of privacy-sensitive context data is not desirable from the privacy viewpoint. However, high quality and quantity of data enable smartness for the AmI, while less and coarse data benefit privacy. This raises a very important problem to the AmI, that is, how to balance the smartness and privacy requirements in an ambient world. In this article, we propose to give to donors the control over the life cycle of their context data, so that users themselves can balance their needs and wishes in terms of smartness and privacy.

1 Introduction

A smart, anticipating, and learning environment will have a great impact on privacy. Ambient Intelligence will be everywhere, is invisible, has powerful sensing capabilities, and most of all has a memory [1]. One of the main difficulties with privacy in the ubiquitous computing, is the way how data is collected. When making a transaction with a web shop, it could be quite clear which kind of data is exchanged. Ubiquitous computing techniques however, such as small sensors, active badges [2], or cameras equipped with powerful image recognizing algorithms, often collect data when people are not aware of it [3,4]. In that case it is possible that people *think* they are in a closed private area (such as coffee rooms), but in *reality* they could be monitored by sensors in that room without being aware of it. This leads to asymetric information [3]. Xiaodong *et al* state that the presence of asymmetric information is the heart of the information privacy problem in ubiquitous computing. In environments with significant asymmetry between the information knowledge of *donor* and *collector*, negative side effects as privacy violations are much harder to overcome.

Several techniques have been proposed in the literature which let donors of the data specify privacy policies, in order to give control about their data to the owners of that data [5,6]. Although such policies are rich enough to let people control who, when, how long, and what kind of information can be disclosed to specific applications, enforcing those policies is usually done through access control. Only relying on access control mechanisms to protect against unauthorized

P. Havinga et al. (Eds.): EUROSSC 2006, LNCS 4272, pp. 255–258, 2006.
© Springer-Verlag Berlin Heidelberg 2006

disclosure of data, is not sufficient enough in terms of privacy protection [7,8]. Perhaps the context databases can be trusted *now*, but they might not be in the future (due to the change of privacy regulation laws for example). Therefore, limited retention techniques are highly desirable to prevent large context histories to be disclosed.

A second problem of traditional privacy policies found in the literature is that they only provide means to express privacy wishes for specific applications. Usage of the data is known in advance, as is the purpose for which the data will be used. Purposes of traditional applications requiring (context) data are atomic in the sense that it is clearly known when purposes are fulfilled or not. For applications which will use context data to learn, infer, and thus to become smarter, it is not clear when such a purpose has been fulfilled, in other words, purposes are *non-atomic*. It is even unclear which services and applications will use the context data in the future.

For static databases containing large datasets with privacy sensitive data (like medical data), anonymization can be used to prevent disclosure of individual privacy sensitive data [9,10,11]. However, anonymization does not always give adequate privacy protection to everyone, and the usability of the data becomes sometimes lower than needed because individual privacy concerns and personal interests are not taken into account. Xiao *et al* [12] recognize this problem and propose to *personalize* the anonymization of privacy sensitive data.

The nature of data used in the ambient smart environments is different and more dynamic than that of traditional static data. The amount of smartness of applications is bound to the quantity and quality of the data they can use. The more accurate the data is, and the more data has been gathered from a certain individual, the better a smart application can learn from that data without user interaction [13]. The challenge is to find the best balance between the quality and quantity of data at the one side, and the privacy sensitivity of the data at the other side.

2   Motivation

Consider a working environment where employees can access the Internet and rank their visited websites. These Internet browsing behaviors are monitored and recorded in a database. Employees can query the database to find interesting websites based on the ranking, and discuss with other employees who have visited and ranked the websites. However, because the starting and end times when an employee made a website visit are also recorded, it is possible to deduce the duration that an employee spends on the Internet per day. Thus, most employees may not want the system to record such sensitive information in the database although the employees do benefit from the offered smart query services. To compromise the smartness and privacy requirements, a self-regulation of sensitive information could be like: one hour/day later, degrade the employee id to his/her group id or even faculty id. In this way, s/he can still use such a query service as *"give me interesting websites visited by the people from the database group last week"*.

3   Approach

We let people (the authors of monitored events) specify *Life-Cycle Policies* which will be bound to the acquired sensitive data [14]. Events are monitored and bound to a context tuple, which could contain *author, location, time et cetera.* This data is stored in a privacy aware context database system, which degrades the data progressively according to the policy. This way, context history can be considered as events (like *a door has been opened*) bound to attributes describing those events. The context values exhibit a certain level of accuracy based on domain generalization graphs. Such generalization graphs together form a n-dimensional space, in which each dimension represents the accuracy of an attribute of the original data tuple. We consider that those levels of accuracy can be classified given the privacy of the information they represent, such that all possible combinations of accuracies form a *n*-dimensional cube. A life-cycle policy can be viewed as a path specification in this cube. Triggered by events (we consider both time and contextual events), the accuracy of context tuples progressively decreases when specified conditions are satisfied.

4   Case Study

A prototype of a system which monitors the Internet browsing behavior of users has been implemented. Websites visited by users will be monitored, enabling smart services like ranking websites, contacting users of the same interests, finding interesting websites visited by members of a certain group, calculation of anonymized statistics, and so on. From the privacy perspective, collecting this data (including times when people were active on the Internet) makes it for malicious parties possible to deduce (for the user) confronting information. Users can specify their life-cycle policies (e.g., degrade time to hour and person id to group after one hour, degrade URL to category after one month, see Figure 1), which are attached with the data and will be stored and executed within a privacy aware context database. By specifying a life-cycle policy, users are sure that data which has been degraded can no longer be misused by malicious parties. Hence, the amount of smartness is reduced to decrease the possibility of misusing the data, consequently increasing privacy.

Fig. 1. A LCP example with states (time, id, url) where ϕ stands for a deleted (or completely degraded) value

Acknowledgments

This work is funded by the Dutch organization for scientific research (NWO-Vidi project) and the Centre for Telematics and Information Technology of the University of Twente.

References

1. Langheinrich, M.: Privacy by design - principles of privacy-aware ubiquitous systems. In: UbiComp '01: Proceedings of the 3rd international conference on Ubiquitous Computing, London, UK, Springer-Verlag (2001) 273–291
2. Want, R., Hopper, A., Falcao, V., Gibbons, J.: The active badge location system. Technical Report 92.1, ORL, 24a Trumpington Street, Cambridge CB2 1QA (1992)
3. Jiang, X., Hong, J.I., Landay, J.A.: Approximate information flows: Socially-based modeling of privacy in ubiquitous computing. In: UbiComp '02: Proceedings of the 4th international conference on Ubiquitous Computing, London, UK, Springer-Verlag (2002) 176–193
4. Little, L., Briggs, P.: Tumult and turmoil: privacy in an ambient world. In: Workshop on Privacy, Trust and Identity Issues for Ambient Intelligence (Pervasive 2006). (2006)
5. W3C: Platform for privacy preferences (P3P) project. http://www.w3.org/P3P/ (2005)
6. Byun, J.W., Bertino, E.: Micro-views, or on how to protect privacy while enhancing data usability. Vision paper CERIAS Tech Report 2005-25, Center for Education and Research in Information Assurance and Security, West Lafayette, IN 47907-2086 (2005)
7. Agrawal, R., Kiernan, J., Srikant, R., Xu, Y.: Hippocratic databases. In: 28th Int'l Conf. on Very Large Databases (VLDB), Hong Kong. (2002)
8. Hong, J.I., Landay, J.A.: An architecture for privacy-sensitive ubiquitous computing. In: MobiSys '04: Proceedings of the 2nd international conference on Mobile systems, applications, and services, New York, NY, USA, ACM Press (2004) 177–189
9. Sweeney, L.: k-anonymity: A model for protecting privacy. International Journal on Uncertainty Fuzziness and Knowledge-based Systems (2002) 557–570
10. Machanavajjhala, A., Gehrke, J., Kifer, D., Venkitasubramaniam, M.: l-diversity: Privacy beyond k-anonymity. In: CLDB. (2006)
11. Chawla, S., Dwork, C., McSherry, F., Smith, A., Wee, H.: Toward privacy in public databases. In: Theory of Cryptography Conference. (2005)
12. Xiao, X., Tao, Y.: Personalized privacy preservation. In: ACM Conference on Management of Data (SIGMOD). (2006)
13. Doom, C.: Get smart: How intelligent technology will enhance our world. Technical report, Computer Sciences Corporation: Leading Edge Forum (2001) A report available from www.csc.com.
14. Anciaux, N., van Heerde, H., Feng, L., Apers, P.: Implanting life-cycle privacy policies in a context database,. Technical Report TR-CTIT-06-03, University of Twente, P.O. Box 217 (2006)

Energy Conservation with EDFI Scheduling

Tjerk Bijlsma and Pierre Jansen

University of Twente, P.O. Box 217, 7500 AE Enschede, The Netherlands
{Bijlsma, Jansen}@cs.utwente.nl

1 Introduction

With the growing popularity of Wireless Sensor Networks (WSNs), the demand increases to perform time critical operations within such networks. A WSN is composed of sensor nodes, which contain a radio, ports for multiple sensors and a microcontroller. These sensor nodes have to provide a wide range of functionality as long as possible, while they use their scarce energy from a battery.

Most of the tasks a sensor node in a WSN has to perform, are periodic tasks. When a sensor node is equipped with a real-time Operating System (OS) it can be guaranteed that these tasks are executed on time.

Since energy is scarce for sensor nodes, the real-time OS can turn unused devices off in idle time to *save energy*. To use the available energy and idle time optimal, the kernel can shutdown the CPU or lower its clock frequency.

A real-time OS should decide how to save energy in the available idle time. The OS needs additional processing power to make such decisions, while processing power and energy are typically scarce for a sensor node. For that reason it is interesting to determine which energy conservation policy is the best.

As state of the art we can mention that multiple energy conservation policies are already available. A distinction can be made between online and offline policies. The *online* energy conserving policies try to scale the voltage and the frequency of the CPU or disables it when idle-time is detected, while the tasks are executed. *Offline* energy conserving policies determine the scaling factor of the task set before it is executed. Both online and offline policies work in combination with Earliest Deadline First (EDF) and Rate Monotonic (RM) scheduling, no shared resources are considered.

2 Policies

Two energy conservation policies based on the *light weight* Earliest Deadline First with Inheritance (EDFI) scheduling are examined. EDFI scheduling is an extension of EDF scheduling that uses inheritance of deadlines in order to enable mutual exclusive access to shared resources. Deadline inheritance allows for straightforward reasoning about feasibility and a simple implementation of the scheduler.

2.1 Temporal Shutdown Scheduling

The online Temporal Shutdown Scheduling (TSS) policy *disables* the microcontroller when idle time is detected. When the scheduler has no running task or

P. Havinga et al. (Eds.): EUROSSC 2006, LNCS 4272, pp. 259–261, 2006.

tasks waiting for execution, it can shutdown the microcontroller. When an interrupt occurs the controller is to be turned on again.

The advantage of this policy is that the task keep their normal lateness. This means that since the tasks are executed at full speed, the time between when they are finished and when they should be finished stays the same. Beside that, a small amount of code is needed to implement the policy.

TSS scheduling has the disadvantage that parts of the microcontroller get disabled. Possibly these parts contained functionality, like fast clocks, that where used by other parts of the sensor node. Some microcontrollers provide caches and registers, that might get lost when the microcontroller is disabled.

2.2   Earliest Deadline First with Inheritance and Scaling

The Earliest Deadline First with Inheritance and Scaling (EDFIS) policy determines a *scaling factor* for the processor frequency. Scaling the frequency causes the tasks to require more execution time. The EDFIS algorithm analyzes the task set in a similar way as the EDF feasibility algorithm of Baruah. Initial the utilization of the task set is chosen as scaling factor for the processor frequency. The deadline events in the task set are examined to verify that the workload of the task set does not exceed the scaled processor capacity, where also the blocking, caused by shared resources, is taken into account. When the workload exceeds the processor capacity the scaling factor is increased, to provide more processor capacity.

EDFIS adjusts the frequency of the microcontroller, keeping all functionality available. The policy performs an analysis when new tasks are inserted in the tasks set, no computations are required at run time.

Note that the scaling possibilities of the EDFIS policy are limited by the amount of blocking and early deadlines of tasks in the task set. Furthermore the scaling causes the tasks to be finished nearer to their deadlines, increasing the lateness of the tasks.

3   Measurements

The energy that is saved by the two policies is examined using the μnode v2.0 of Ambient Systems, which is equipped with a MSP430F1611 microcontroller. The best-case and worst-case situations are examined for both policies.

The tests show that at a utilization of 0.10 the power saved by the TSS policy lies between the 31% and 42%. The difference is mainly caused by the frequency at which the scheduler is called, which is related to the period of the tasks in the task set. In the worst-case situation EDFIS can not scale the task set due to small deadlines or blocking. The maximum amount of power saved by EDFIS is 34% at a utilization of 0.10. Another interesting result shown in the tests is that EDFIS outperforms TSS when the utilization of the tasks set is above 0.60.

4   Conclusion

Not a single solution can be provided the best power savings on the μnode v2.0. In case the utilization of the task set is low, typically below 0.40, the TSS policy can deliver power savings up to 42%. Drawback of this solution is that devices cannot use the fast timers. When the utilization is above 0.60 the EDFIS policy shows a larger amount of saved power. Restriction is that the task set should not have to much blocking or small deadlines.

RuleCaster: A Programming System for Wireless Sensor Networks

Urs Bischoff and Gerd Kortuem

Computing Department
Lancaster University
Lancaster, LA1 4WA, UK
u.bischoff@lancaster.ac.uk, kortuem@comp.lancs.ac.uk

1 Introduction

Writing and deploying software can be a challenging task. Especially when dealing with a wireless sensor network (WSN) consisting of a large number of nodes. Dealing with each node individually is not feasible. It is too time consuming, too costly and too error prone. This is even more relevant if we consider a dynamic environment with changing device configurations and often changing requirements. What we need are suitable methods for designing, implementing and deploying applications.

We argue that instead of focusing on each individual node independently we need to focus on programming the entire network as a whole. Knowing the individual node that executes the aplication is of less interest. We need to find ways to define applications independently of the underlying distributed computing environment and a way to inject them dynamically into the network.

2 Programming the Network

We are developing a programming system for WSNs (RuleCaster) that addresses the network as a whole. It consists of four components: (1) a high-level programming language, (2) a dynamic model of the WSN describing the capabilities of its nodes, (3) a compiler that splits the application written in the high-level language into several tasks and assigns them to the nodes, and (4) a run-time system on the nodes that can execute the tasks.

The programmer writes an application for the network in the high-level programming language. The compiler uses the information given by the network model to split the application into different tasks; it then assigns these tasks to suitable nodes in the network. Each task is a program executed by one node. This process is illustrated in Fig. 1. The tasks prescribe computation done by the nodes as well as how they have to collaborate in order to get additional information needed to execute it.

3 Example

Let us assume a storage facility with a large number of different sensor nodes embedded in the environment and the storage objects. We want to use this

P. Havinga et al. (Eds.): EUROSSC 2006, LNCS 4272, pp. 262–263, 2006.

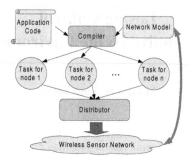

Fig. 1. A global application definition is translated into a distributed application running in the network

WSN to detect situations that require to take certain actions. We want to know if incompatible products are stored in proximity, for example. Or we want to detect if products are in a room that is too hot for them. Or situations where a product is in an unapproved area for too long should be prevented.

The programmer uses the high-level language to write the application rules for the whole storage facility network. Then, a node is automatically assigned a task that represents its contribution towards the whole storage facility application. Each node has different capabilities: some can measure temperature, others can measure distances and detect the presence of other nearby objects, for example. In order to execute certain tasks, a node has to collaborate with other nodes. A node that knows that it is attached to an object that has to be kept cool should be able to decide whether the environmental conditions are suitable. If it does not have a temperature sensor it has to collaborate with a nearby node to execute this task. All this information is encoded in the individual task descriptions that can be executed by the nodes.

4   Ongoing and Future Work

We have defined a high-level language that is based on rules. The declarative nature of this rule-based language is essential because it allows us to seperate application logic from application execution. It does not specify where and how the application is executed in the network; it only specifies what the network as a whole has to do and what results the user expects.

We have developed the RuleCaster compiler that splits the application into tasks. Tasks are described in form of an intermediate language. They are executed by a runtime system on wireless sensor nodes.

It is the RuleCaster compiler that can decide how an application is split up and distributed in the network. Current research investigates different distribution strategies. A suitable distribution strategy can depend on application requirements or the characteristics of the network. Several strategies are possible: minimal energy consumption or minimal communication are two examples.

Losing Control
in Pro-active Home Environments

Martijn H. Vastenburg

Delft University of Technology, Faculty of Industrial Design Engineering
Landbergstraat 15, 2628 CE, Delft, The Netherlands
m.h.vastenburg@tudelft.nl

Abstract. Although context-aware and autonomous services could eventually automatically adapt home environments to user needs, there is a risk of creating environments in which people experience a lack of control. The perceived level of user control in aware home environments might be correlated to the level of system initiative. Consequently, perceived user control might be improved by dynamically adapting the system-initiative level to the individual users and context. Towards creating the optimal balance between system-initiative and user control in a sensitive and personalized home environment, an experimental study has been conducted in a realistic setting. In the experiment, participants were willing to give up partial control. However, participants regularly switched to manual control in order to slightly change system settings. The study suggests a feedback mechanism is crucial to improve system behavior in time, and increase the acceptability of autonomous services in the home.

Keywords: mixed initiative interaction, context-aware homes, personalization.

1 Introduction

To create an environment suitable to a task or range of activities, users are typically required to adjust device settings manually. To control a small number of devices, manual adjustment may not take much effort. However, the number of devices in the home is growing, as well as the number of functions per device. If a new button were added for every new function in a home control system, users would most likely be overwhelmed by system complexity. Ubiquitous computing technology enables the environment to pro-actively adapt the surroundings to user state [5], activities [1, 2] or domestic routines [4].

Although context-aware and autonomous services could eventually automatically adapt to the user needs, there is a risk of creating an environment in which people experience a lack of control. In an experimental case study [3], three approaches for interaction between humans and context-aware mobile applications have been compared: personalization, passive context-awareness and active context-awareness. The study showed that users feel less in control when using either passive or active context-aware applications as compared to personalizing applications. However, users preferred the context-aware applications. Apparently, if the gain in usefulness is big

P. Havinga et al. (Eds.): EUROSSC 2006, LNCS 4272, pp. 264 – 265, 2006.

enough, people are willing to give up partial control. The applicability of these findings in the living room environment has been studied in the present experiment.

This poster describes a user study in which perceived user control of aware home systems is measured. The research goal was to find out how much initiative people are willing to delegate to an aware home system. Eventually, this knowledge will be used to create adaptivity models and improve user-system interaction.

2 User Study

The user study was located in StudioHome, a living room laboratory. Ten couples of subjects (10 women, 10 men) participated in the experiment. Each couple participated for one evening. Participants were asked to perform user activities they normally do at home.

A central touch-screen based application could be used to control the lights, music and atmosphere. The home control application could be used in three modes: manual, passive context awareness (system suggests changes), and active context awareness (autonomous changes). Participants were asked to use the application to adapt the room to their needs. Using a questionnaire, participants were asked to rate the perceived user control before and after each user activity.

3 Reflections on the Design and Approach

In general, participants were positive on pro-active system behavior. In the living room lab setting, they were able to experience the system for three hours. Inter-subject data shows the perceived level of user control increases in time. However, participants indicated their dislike of some of the pre-programmed settings. Consequently, they disliked the system actions, resulting in a decrease in perceived user control. For future studies, a feedback mechanism is needed to rapidly adapt models of user preferences to the individual needs.

References

1. Abowd GD, Mynatt ED (2000) Charting past, present, and future research in ubiquitous computing. ACM Trans. Comput.-Hum. Interact., 7(1): 29-58
2. Bardram JE, Christensen HB (2004) Open Issues in Activity-Based and Task-Level Computing. In First International Workshop on Computer Support for Human Tasks and Activities, Vienna, Austria, 2004. Centre for Pervasive Computing Technical Reports
3. Barkhuus L, Dey A (2003) Is Context-Aware Computing Taking Control Away from the User? Three Levels of Interactivity Examined (pp. 1-8)
4. Tolmie P, Pycock J, Diggins T, MacLean A, Karsenty A (2002) Unremarkable Computing. CHI Letters, 1(1): 399-406
5. Vastenburg MH, Keyson DV, de Ridder H (2004) Interrupting People at Home. In IEEE International Conference on Systems, Man and Cybernetics, The Hague, The Netherlands, 2004

Author Index

An, Xiangdong   159
Anawar, Syarulnaziah   218
Anciaux, Nicolas   255
Apers, Peter M.G.   255

Baggio, Aline   39
Bałos, Kazimierz   54
Bijlsma, Tjerk   259
Bischoff, Urs   262
Blackstock, Michael   113
Broens, Tom   82
Brown, James   248

Cemgil, A. Taylan   240
Cercone, Nick   159
Chen, Ke-An   204
Cho, Joonmyun   174
Cho, Seong-Yun   222
Choi, Young   25
Cottingham, David N.   226
Coyle, Lorcan   218

Davies, Jonathan J.   226
Ditzel, Maarten   15
Dobson, Simon   218, 238
Dulman, Stefan   1
Durmaz Incel, Ozlem   1

Feng, Ling   255
Finney, Joe   248

Halkes, Gertjan P.   39
Halteren, Aart van   82
Han, Kijun   25
Han, Su-Young   222
Harle, R.K.   128
Heerde, Harold van   255
Hesselman, Cristian   67
Hong, Chung-Seong   174

Iacob, Sorin   67

Jansen, Pierre   1, 259
Janssen, Johan   15
Jones, Brian D.   226

Jung, Eui-Hyun   222
Jutla, Dawn   159

Kao, Kuo-Fong   204
Kim, Hyun   174
Kim, Hyunsook   25
Kim, Kyungmi   25
Kim, Sungrim   230
Kim, Yong-Pyo   222
Konstantas, Dimitri   250
Kortuem, Gerd   248, 262
Krasic, Charles   113
Kröse, Ben J.A.   240
Kwon, Joonhee   230

Lageweg, Caspar   15
Langendoen, Koen G.   39
Lea, Rodger   113
Lee, Hyun-Chan   174
Lee, Jae Sik   190
Lee, Jin Chun   190
Lee, Kang-Woo   174
Lee, Sukgyu   25
Liao, I-En   204
Li, Jun   143
Lowton, Mark   248
Lu, Jian   143
Lu, Wentian   143

Ma, Rui   244
Ma, Xiaoxing   143

Neely, Steve   238
Neisse, Richardo   234
Nixon, Paddy   218, 238

O'Grady, Michael J.   244
O'Hare, Gregory M.P.   244

Park, Yong-Jin   222
Pawar, Pravin   67

Rarau, Anca   98

Salomie, Ioan   98
Sinderen, Marten van   82, 234

Stevenson, Graeme   238
Suh, Young-Ho   174
Szydło, Tomasz   54
Szymacha, Robert   54

Tao, Xianping   143
Tokmakoff, Andrew   67

Vastenburg, Martijn H.   264
Villalba, Alfredo   250

Wegdam, Maarten   234
Williamson, Graham   238

Zajdel, Wojciech   240
Zieliński, Krzysztof   54

Lecture Notes in Computer Science

For information about Vols. 1–4175

please contact your bookseller or Springer

Vol. 4272: P. Havinga, M. Lijding, N. Meratnia, M. Wegdam (Eds.), Smart Sensing and Context. XI, 267 pages. 2006.

Vol. 4270: H. Zha, Z. Pan, H. Thwaites, A.C. Addison, M. Forte (Eds.), Interactive Technologies and Sociotechnical Systems. XVI, 547 pages. 2006.

Vol. 4269: R. State, S. van der Meer, D. O'Sullivan, T. Pfeifer (Eds.), Large Scale Management of Distributed Systems. XIII, 282 pages. 2006.

Vol. 4267: A. Helmy, B. Jennings, L. Murphy, T. Pfeifer (Eds.), Autonomic Management of Mobile Multimedia Services. XIII, 257 pages. 2006.

Vol. 4265: N. Lavrač, L. Todorovski, K.P. Jantke (Eds.), Discovery Science. XIV, 384 pages. 2006. (Sublibrary LNAI).

Vol. 4264: J.L. Balcázar, P.M. Long, F. Stephan (Eds.), Algorithmic Learning Theory. XIII, 393 pages. 2006. (Sublibrary LNAI).

Vol. 4257: I. Richardson, P. Runeson, R. Messnarz (Eds.), Software Process Improvement. XI, 219 pages. 2006.

Vol. 4254: T. Grust, H. Höpfner, A. Illarramendi, S. Jablonski, M. Mesiti, S. Müller, P.-L. Patranjan, K.-U. Sattler, M. Spiliopoulou (Eds.), Current Trends in Database Technology – EDBT 2006. XXXI, 932 pages. 2006.

Vol. 4253: B. Gabrys, R.J. Howlett, L.C. Jain (Eds.), Knowledge-Based Intelligent Information and Engineering Systems, Part III. XXXII, 1301 pages. 2006. (Sublibrary LNAI).

Vol. 4252: B. Gabrys, R.J. Howlett, L.C. Jain (Eds.), Knowledge-Based Intelligent Information and Engineering Systems, Part II. XXXIII, 1335 pages. 2006. (Sublibrary LNAI).

Vol. 4251: B. Gabrys, R.J. Howlett, L.C. Jain (Eds.), Knowledge-Based Intelligent Information and Engineering Systems, Part I. LXVI, 1297 pages. 2006. (Sublibrary LNAI).

Vol. 4249: L. Goubin, M. Matsui (Eds.), Cryptographic Hardware and Embedded Systems - CHES 2006. XII, 462 pages. 2006.

Vol. 4248: S. Staab, V. Svátek (Eds.), Engineering Knowledge in the Age of the Semantic Web. XIV, 400 pages. 2006. (Sublibrary LNAI).

Vol. 4247: T.-D. Wang, X. Li, S.-H. Chen, X. Wang, H. Abbass, H. Iba, G. Chen, X. Yao (Eds.), Simulated Evolution and Learning. XXI, 940 pages. 2006.

Vol. 4245: A. Kuba, L.G. Nyúl, K. Palágyi (Eds.), Discrete Geometry for Computer Imagery. XIII, 688 pages. 2006.

Vol. 4243: T. Yakhno, E.J. Neuhold (Eds.), Advances in Information Systems. XIII, 420 pages. 2006.

Vol. 4241: R.R. Beichel, M. Sonka (Eds.), Computer Vision Approaches to Medical Image Analysis. XI, 262 pages. 2006.

Vol. 4239: H.Y. Youn, M. Kim, H. Morikawa (Eds.), Ubiquitous Computing Systems. XVI, 548 pages. 2006.

Vol. 4238: Y.-T. Kim, M. Takano (Eds.), Management of Convergence Networks and Services. XVIII, 605 pages. 2006.

Vol. 4236: L. Breveglieri, I. Koren, D. Naccache, J.-P. Seifert (Eds.), Fault Diagnosis and Tolerance in Cryptography. XIII, 253 pages. 2006.

Vol. 4234: I. King, J. Wang, L. Chan, D. Wang (Eds.), Neural Information Processing, Part III. XXII, 1227 pages. 2006.

Vol. 4233: I. King, J. Wang, L. Chan, D. Wang (Eds.), Neural Information Processing, Part II. XXII, 1203 pages. 2006.

Vol. 4232: I. King, J. Wang, L. Chan, D. Wang (Eds.), Neural Information Processing, Part I. XLVI, 1153 pages. 2006.

Vol. 4229: E. Najm, J.F. Pradat-Peyre, V.V. Donzeau-Gouge (Eds.), Formal Techniques for Networked and Distributed Systems - FORTE 2006. X, 486 pages. 2006.

Vol. 4228: D.E. Lightfoot, C.A. Szyperski (Eds.), Modular Programming Languages. X, 415 pages. 2006.

Vol. 4227: W. Nejdl, K. Tochtermann (Eds.), Innovative Approaches for Learning and Knowledge Sharing. XVII, 721 pages. 2006.

Vol. 4225: J.F. Martínez-Trinidad, J.A. Carrasco Ochoa, J. Kittler (Eds.), Progress in Pattern Recognition, Image Analysis and Applications. XIX, 995 pages. 2006.

Vol. 4224: E. Corchado, H. Yin, V. Botti, C. Fyfe (Eds.), Intelligent Data Engineering and Automated Learning – IDEAL 2006. XXVII, 1447 pages. 2006.

Vol. 4223: L. Wang, L. Jiao, G. Shi, X. Li, J. Liu (Eds.), Fuzzy Systems and Knowledge Discovery. XXVIII, 1335 pages. 2006. (Sublibrary LNAI).

Vol. 4222: L. Jiao, L. Wang, X. Gao, J. Liu, F. Wu (Eds.), Advances in Natural Computation, Part II. XLII, 998 pages. 2006.

Vol. 4221: L. Jiao, L. Wang, X. Gao, J. Liu, F. Wu (Eds.), Advances in Natural Computation, Part I. XLI, 992 pages. 2006.

Vol. 4219: D. Zamboni, C. Kruegel (Eds.), Recent Advances in Intrusion Detection. XII, 331 pages. 2006.

Vol. 4218: S. Graf, W. Zhang (Eds.), Automated Technology for Verification and Analysis. XIV, 540 pages. 2006.

Vol. 4217: P. Cuenca, L. Orozco-Barbosa (Eds.), Personal Wireless Communications. XV, 532 pages. 2006.

Vol. 4216: M.R. Berthold, R. Glen, I. Fischer (Eds.), Computational Life Sciences II. XIII, 269 pages. 2006. (Sublibrary LNBI).

Vol. 4215: D.W. Embley, A. Olivé, S. Ram (Eds.), Conceptual Modeling - ER 2006. XVI, 590 pages. 2006.

Vol. 4213: J. Fürnkranz, T. Scheffer, M. Spiliopoulou (Eds.), Knowledge Discovery in Databases: PKDD 2006. XXII, 660 pages. 2006. (Sublibrary LNAI).

Vol. 4212: J. Fürnkranz, T. Scheffer, M. Spiliopoulou (Eds.), Machine Learning: ECML 2006. XXIII, 851 pages. 2006. (Sublibrary LNAI).

Vol. 4211: P. Vogt, Y. Sugita, E. Tuci, C. Nehaniv (Eds.), Symbol Grounding and Beyond. VIII, 237 pages. 2006. (Sublibrary LNAI).

Vol. 4210: C. Priami (Ed.), Computational Methods in Systems Biology. X, 323 pages. 2006. (Sublibrary LNBI).

Vol. 4209: F. Crestani, P. Ferragina, M. Sanderson (Eds.), String Processing and Information Retrieval. XIV, 367 pages. 2006.

Vol. 4208: M. Gerndt, D. Kranzlmüller (Eds.), High Performance Computing and Communications. XXII, 938 pages. 2006.

Vol. 4207: Z. Ésik (Ed.), Computer Science Logic. XII, 627 pages. 2006.

Vol. 4206: P. Dourish, A. Friday (Eds.), UbiComp 2006: Ubiquitous Computing. XIX, 526 pages. 2006.

Vol. 4205: G. Bourque, N. El-Mabrouk (Eds.), Comparative Genomics. X, 231 pages. 2006. (Sublibrary LNBI).

Vol. 4204: F. Benhamou (Ed.), Principles and Practice of Constraint Programming - CP 2006. XVIII, 774 pages. 2006.

Vol. 4203: F. Esposito, Z.W. Raś, D. Malerba, G. Semeraro (Eds.), Foundations of Intelligent Systems. XVIII, 767 pages. 2006. (Sublibrary LNAI).

Vol. 4202: E. Asarin, P. Bouyer (Eds.), Formal Modeling and Analysis of Timed Systems. XI, 369 pages. 2006.

Vol. 4201: Y. Sakakibara, S. Kobayashi, K. Sato, T. Nishino, E. Tomita (Eds.), Grammatical Inference: Algorithms and Applications. XII, 359 pages. 2006. (Sublibrary LNAI).

Vol. 4200: I.F.C. Smith (Ed.), Intelligent Computing in Engineering and Architecture. XIII, 692 pages. 2006. (Sublibrary LNAI).

Vol. 4199: O. Nierstrasz, J. Whittle, D. Harel, G. Reggio (Eds.), Model Driven Engineering Languages and Systems. XVI, 798 pages. 2006.

Vol. 4198: O. Nasraoui, O. Zaiane, M. Spiliopoulou, B. Mobasher, B. Masand, P. Yu (Eds.), Advances in Web Minding and Web Usage Analysis. IX, 177 pages. 2006. (Sublibrary LNAI).

Vol. 4197: M. Raubal, H.J. Miller, A.U. Frank, M.F. Goodchild (Eds.), Geographic, Information Science. XIII, 419 pages. 2006.

Vol. 4196: K. Fischer, I.J. Timm, E. André, N. Zhong (Eds.), Multiagent System Technologies. X, 185 pages. 2006. (Sublibrary LNAI).

Vol. 4195: D. Gaiti, G. Pujolle, E. Al-Shaer, K. Calvert, S. Dobson, G. Leduc, O. Martikainen (Eds.), Autonomic Networking. IX, 316 pages. 2006.

Vol. 4194: V.G. Ganzha, E.W. Mayr, E.V. Vorozhtsov (Eds.), Computer Algebra in Scientific Computing. XI, 313 pages. 2006.

Vol. 4193: T.P. Runarsson, H.-G. Beyer, E. Burke, J.J. Merelo-Guervós, L.D. Whitley, X. Yao (Eds.), Parallel Problem Solving from Nature - PPSN IX. XIX, 1061 pages. 2006.

Vol. 4192: B. Mohr, J.L. Träff, J. Worringen, J. Dongarra (Eds.), Recent Advances in Parallel Virtual Machine and Message Passing Interface. XVI, 414 pages. 2006.

Vol. 4191: R. Larsen, M. Nielsen, J. Sporring (Eds.), Medical Image Computing and Computer-Assisted Intervention – MICCAI 2006, Part II. XXXVIII, 981 pages. 2006.

Vol. 4190: R. Larsen, M. Nielsen, J. Sporring (Eds.), Medical Image Computing and Computer-Assisted Intervention – MICCAI 2006, Part I. XXXVVIII, 949 pages. 2006.

Vol. 4189: D. Gollmann, J. Meier, A. Sabelfeld (Eds.), Computer Security – ESORICS 2006. XI, 548 pages. 2006.

Vol. 4188: P. Sojka, I. Kopeček, K. Pala (Eds.), Text, Speech and Dialogue. XV, 721 pages. 2006. (Sublibrary LNAI).

Vol. 4187: J.J. Alferes, J. Bailey, W. May, U. Schwertel (Eds.), Principles and Practice of Semantic Web Reasoning. XI, 277 pages. 2006.

Vol. 4186: C. Jesshope, C. Egan (Eds.), Advances in Computer Systems Architecture. XIV, 605 pages. 2006.

Vol. 4185: R. Mizoguchi, Z. Shi, F. Giunchiglia (Eds.), The Semantic Web – ASWC 2006. XX, 778 pages. 2006.

Vol. 4184: M. Bravetti, M. Núñez, G. Zavattaro (Eds.), Web Services and Formal Methods. X, 289 pages. 2006.

Vol. 4183: J. Euzenat, J. Domingue (Eds.), Artificial Intelligence: Methodology, Systems, and Applications. XIII, 291 pages. 2006. (Sublibrary LNAI).

Vol. 4182: H.T. Ng, M.-K. Leong, M.-Y. Kan, D. Ji (Eds.), Information Retrieval Technology. XVI, 684 pages. 2006.

Vol. 4180: M. Kohlhase, OMDoc – An Open Markup Format for Mathematical Documents [version 1.2]. XIX, 428 pages. 2006. (Sublibrary LNAI).

Vol. 4179: J. Blanc-Talon, W. Philips, D. Popescu, P. Scheunders (Eds.), Advanced Concepts for Intelligent Vision Systems. XXIV, 1224 pages. 2006.

Vol. 4178: A. Corradini, H. Ehrig, U. Montanari, L. Ribeiro, G. Rozenberg (Eds.), Graph Transformations. XII, 473 pages. 2006.

Vol. 4177: R. Marín, E. Onaindía, A. Bugarín, J. Santos (Eds.), Current Topics in Artificial Intelligence. XV, 482 pages. 2006. (Sublibrary LNAI).

Vol. 4176: S.K. Katsikas, J. Lopez, M. Backes, S. Gritzalis, B. Preneel (Eds.), Information Security. XIV, 548 pages. 2006.